Praise for the authors' previous work, *Safety Man* *Approach*:

"…a much needed and overdue academic contribution to the field of safety management."
—*Ergonomics*

"…some excellent insights and arguments…very valuable reading…"
—*Ergonomics*

"…a new perspective…distinctive…a unique resource worth buying."
—*E-Streams* (*Electronic reviews of Science & Technology References covering Engineering, Agriculture, Medicine, and Science*)

"…this book truly presents a philosophical challenge to the precepts held in the disciplines of safety management and occupational psychology."
—*E-Streams*

"the…authors…have developed a new view of safety management…"
—*Safety and Health Practitioner*

"…essential reading for an up-to-date perspective on near miss reporting and safety management."
—*Risk Analysis*

BEYOND HUMAN ERROR

TAXONOMIES AND SAFETY SCIENCE

Brendan Wallace
University of Glasgow
Glasgow, Scotland, U.K.

Alastair Ross
University of Strathclyde
Glasgow, Scotland, U.K.

CRC Press
Taylor & Francis Group
Boca Raton London New York

CRC Press is an imprint of the
Taylor & Francis Group, an **informa** business

A TAYLOR & FRANCIS BOOK

CRC Press
Taylor & Francis Group
6000 Broken Sound Parkway NW, Suite 300
Boca Raton, FL 33487-2742

First issued in paperback 2019

© 2006 by Taylor & Francis Group, LLC
CRC Press is an imprint of Taylor & Francis Group, an Informa business

No claim to original U.S. Government works

ISBN-13: 978-0-8493-2718-6 (hbk)
ISBN-13: 978-0-367-39103-4 (pbk)

Library of Congress Card Number 2005057495

Library of Congress Cataloging-in-Publication Data

Wallace, Brendan.
   Beyond human error : taxonomies and safety science / Brendan Wallace, Alastair Ross.
       p. cm.
   Includes bibliographical references and index.
   ISBN 0-8493-2718-0 (acid-free paper)
   1. Accidents--Research--Methodology. 2. Accidents--Prevention. 3. Science--Philosophy. I. Ross, Alastair, 1966- II. Title.

HV675.W33 2006
363.1001--dc22   2005057495

Visit the Taylor & Francis Web site at
http://www.taylorandfrancis.com

and the CRC Press Web site at
http://www.crcpress.com

Dedication

This book is dedicated to Mike and Eileen Wallace, Lorna Clark, Lesley Abernethy, and Ian and Ann Ross.

Foreword

There is an extensive body of literature on safety management and an equally impressive number of safety management "systems" available to anyone with a problem and a bottomless wallet. The literature is dominated largely by "how-to-do-it" texts, and the systems offered for sale are generally of the "here's something you can do it with" type. What is lacking in the safety management arena are texts that seek to answer the question "How do to what?" (there are a few notable exceptions to this generalization) and methods that do justice to the subtleties required when compiling taxonomies, when looking for root causes, when defining *knowledge*, and when exploring the no-man's-land that lies between obvious technical failures and obvious human incompetence, where the two are intricately intertwined.

As a rule, technical failures and design problems involve human factors somewhere, and invariably human error manifests within a specific technical environment. Keeping these two aspects of work apart and ticking a box called "equipment" or a box called "human error" leads to unhelpful and misleading data that at best might serve paperwork functions in connection with site licenses or safety cases but often, in our view, fail to translate into practical actions other than applying cautionary notices bearing banal messages about paying attention and using care to walls.

In a previous book, *Safety Management: A Qualitative Systems Approach*, our team tried to address some of these issues and to present a text that covered the philosophy of human action, the psychology and economics of causality, and the practical problems caused by flawed taxonomies and unreliable coding. In particular, we attempted to dispel the often-mythical beliefs surrounding general assumptions about the assumed clear difference between objective and subjective knowledge and to indicate that the way out of the dilemma is to insist on consensus. We also tried to point out the value that can be added to technical and human factors databases by a principled statistical approach to the analysis of natural (in-your-own-words) incident reports.

Now, Brendan Wallace and Alastair Ross attempt to take these arguments a step further in the current book. The message is that the search for pragmatic and practical solutions to safety management problems requires a grasp of certain elementary philosophies about causality, about cognition, and about how human beings categorize the world; a grasp of some basic statistical procedures that can be utilized to assess the reliability of categorization and coding processes; a more sophisticated view of prediction from data; and a meaningful grasp of what constitutes a "significant" finding. Above all, the message is that effective safety management requires an understanding of the dialectics that lie at the heart of human behavior and thinking and a willingness to take into account the primacy of subjectivity, motives, and intentions. Although for certain purposes people may be conceptualized as biological systems, information processors, or digital computers, they are not, at the end of the day, machines. They may be like machines in certain respects, but machines do not have motives and intentions. The purposiveness of machine systems is entirely a consequence of human purposiveness, and that is where the search for safe and efficient systems of work must start.

Professor John B. Davies
Department of Psychology
University of Strathclyde
Glasgow, United Kingdom

Authors

Dr. Brendan Wallace is a Research Fellow in the Sociology Department at the University of Glasgow, United Kingdom. He is the author of *Safety Management: A Qualitative Systems Approach* (with Alastair Ross, John Davies, and Linda Wright) and *External-to-Vehicle Driver Distraction*. He has carried out research in taxonomic and safety issues for the nuclear industry, the railways, and the police force. His current interests include modern philosophy of mind, new approaches in psychology, and the sociology of risk (especially as this affects young people).

Dr. Alastair Ross is a Senior Research Fellow at the Centre for Applied Social Psychology at the University of Strathclyde, Glasgow, United Kingdom, and a Senior Consultant for Human Factors Analysts Limited. He has worked with taxonomies in nuclear, rail, public health, and military contexts. His current interests include attribution, taxonomic arrangement, and functional discourse.

Acknowledgments

We would like to thank staff at the University of Strathclyde and at Human Factors Analysts Limited, Glasgow, who worked on the CIRAS (Confidential Incident Reporting and Analysis System) project (especially Megan Sudbery, Deborah Grossman, and Rachel Macdonald) and Jim Baxter, Matthew White, and James Harris for early work on SECAS (Strathclyde Event Coding and Analysis System).

We are also indebted to many people in industry and other areas of public life who have collaborated with us on safety and risk-related projects. Particular thanks are given to Andrew Fraser at Central Scotland Roads Accident Investigation Unit; Iain Carrick and Dave Howie at British Energy; the staff at Iarnród Éireann in Ireland involved in CARA (Confidential Accident Reporting and Analysis); all United Kingdom rail companies that volunteered for CIRAS before it was mandated nationally (Scotrail, Great North Eastern Railways, Virgin Trains, First Engineering, G. T. Rail Maintenance, Railtrack Scotland Zone); other organizations we have been involved with in taxonomic work (including Rolls-Royce PLC, BAE Systems, NHS Quality Improvement Scotland); and all our "academic" friends and confidants (especially Tony Anderson, Derek Heim, and Malcolm Hill).

We particularly thank Mike Wallace and Megan Sudbery for reading versions of this text and John Davies for many valuable suggestions and for writing the foreword.

Introduction

This book is about the theory and practice of safety science. Specifically, it argues that many of the current practices in the field are fundamentally misguided, and that only a radical revision of the fundamental philosophy of the area will lead to success in what practitioners and researchers are supposed to be doing: reducing accidents and promoting safety.

This is a large claim, and as a result this book covers a far wider range than most equivalent books in safety science, moving from statistics, to social psychology, to the theory and practice of reporting systems, and so on. However, there is a general thrust to the book that we hope is clear and unites all the various themes and topics covered. As a general theme, we argue that safety science has been misled into trying to model itself on the "hard" sciences (specifically physics). There are two ways in which we could have gone about presenting this thesis. In our last book (*Safety Management: A Qualitative Systems Approach*), we argued that the human sciences are fundamentally different from the hard sciences, and that the human sciences (of which safety science is a part) should develop their own theories and methodologies. In this book, we take a more radical line. Now, we argue that even the hard sciences are not, in fact, as hard as they make themselves out to be, and that the mistake safety scientists (and by implication psychologists and sociologists) have made is to model their science on a concept of the hard sciences that does not really exist. In the last 20 years, work by sociologists of science and philosophers has demonstrated that the "popular image" of physics and chemistry (of neutral objective scientists discovering timeless laws of nature) is not sustainable. Rather, it has been demonstrated that scientists have biases and prejudices like everyone else; that there are influences of gender, ethnicity, and power; and that there really is no single scientific method that applies at all times and to all situations.

However, even though we are now pushing this more radical line, the implications are much the same. Safety science should aim to be an autonomous science that develops its own methodologies and techniques (this applies equally to sociology, psychology, economics, and the rest) and not one that simply apes the methods of physics and chemistry (or, to be more specific, physics and chemistry as seen by nonphysicists and nonchemists).

This emphasis in favor of the particular versus the general has informed our work. So, whereas in Chapter 1 we lay out the general philosophy of the book, in Chapter 2 we look at our specific projects, setting up reporting systems for the British railway and nuclear industries. Apart from the emphasis on reliability/consensus trials (an essential part of this process, we argue) the key point of this chapter is that such projects are and must be situation/organization specific. In other words, there cannot be a single "off-the-shelf," "one-size-fits-all" taxonomy (e.g., to describe "human error") without defining the context.

In Chapters 3 and 4, we look at output from such systems. Again, apart from the specifics of how we deal with data and analyze them, there is a general theme running through these two chapters: our critique of "frequentist" statistics. We argue that frequentism is the "statistical" side (as it were) of what we (following Donald Schön) term *technical rationality* (TR): the view of science first stated by Galileo and Descartes. Instead, we champion Bayesian and other methodologies that do not lead us into the same (false) assumptions as frequentism.

In Chapter 5, we look at psychology, specifically the information processing/cognitivist approach that has been common since 1956. Again, we argue that the information processing view of cognition relies on various assumptions that date to Descartes, assumptions that can and should be overturned if safety science (and psychology) is to progress. Chapter 6 looks at systems theory and is proposed as a way of looking at complex sociotechnical events. In Chapter 7, we put this technique to work by analyzing the *Challenger* and *Columbia* disasters; Chapter 8 examines the main themes of the role of rules and of self-organization and empowerment. We contrast two views of an organization, a top-down, rule-governed view and a bottom-up, emergentist view. We argue that it is this latter view that should be the view of the safety manager, and that the safety manager's task is to facilitate the organization to self-organize. Safety science is (or should be) the various techniques developed to enable the safety manager to do this.

To conclude, this book argues that safety science has taken over the philosophy of TR as its guiding framework. But, TR is not the only philosophy of science available. There are others. Frequentism in statistics, cognitivism, and reductionism are examples of the use of TR in safety science.

TR can (and should) be criticized. It has worked in the past, but as safety science develops, it is likely that TR will prove to be an increasingly inadequate methodology and philosophy. Instead of TR, we propose a view of safety science as a taxonomic science, the use of a broadly Bayesian framework in terms of the statistical analysis of data, and the view of the safety manager/scientist as a facilitator (rather than an objective, technocratic expert). Finally, in the conclusion we argue that this book has implications that reach beyond "pure" safety science, and we describe very briefly some projects in fields as diverse as criminology and social work in which this approach has been beneficial. Therefore, we conclude by arguing that this approach has implications for the social sciences as a whole.

These are large and controversial claims. However, they are based on years of work and numerous projects in various high-consequence industries. Moreover, while the goals of this "new" safety science are more modest than those of the old, they might at least be achievable, whereas, we will argue, the project of the old safety science (to turn safety science into a "rigorous" science like physics) was always doomed to failure.

A GUIDE FOR THE READER

When one of us (B.W.) mentioned the theme of this book to a colleague, he was asked an interesting question: "Is your book a book of theory or a practical book?" It is interesting that this was the first question that sprang to mind as the gap between theory and practice and the way this dichotomy reproduces itself throughout the Western education system is one of the themes of this book. And, what can we say? This is a book of practical solutions, but therefore (we would argue) it is also a book of theory: how could it be otherwise?

Unfortunately, this dichotomy is still a popular one. So, in fairness to the reader, we should point out that those more interested in actually building a reporting system (i.e., the practical stuff) should probably start at Chapter 3 and then go on to Chapter 4. The appendix contains the "hard stuff" about how to carry out a reliability trial. For those who are more interested in taxonomy theory in general (e.g., as an alternative to the various qualitative methodologies available currently in the social sciences), Chapter 2 is probably the best place to start. Those more interested in the general theory that lies behind our approach, on the other hand, will probably want to read the whole book, although they may find that close reading of Chapters 3 and 4 is not really necessary as long as they grasp the basic points about frequentism and the importance of reliability trials.

And now, to begin at the beginning!

Contents

1 Safety and Science

Given the rapid growth in the number of books on error, safety, and safety management in recent years, the long-suffering reader will doubtless be asking first: why another book on safety? It should be explained at the outset that this book in some ways covers similar ground as other books in the field. It discusses accidents, why they happen, how they can be prevented, and related issues. However, in our defense, we would argue that in some important respects we have a different view of the subject from most safety experts. Partly this is because of our background (in social psychology, not engineering). However, the way we have come to look at the discipline is also partly because of the nature of our experiences in this type of work. Specifically, it stems from a gradual disillusionment we have felt as we have tried (and failed) to apply the standard methodologies of academic safety science to real-world safety situations. In our opinion, we are not alone in this disillusionment. Off the record, we have also heard many safety practitioners be overtly dismissive about the value of academic texts (unfortunately, texts written by psychologists seem to be viewed with particular disdain, we have to admit with chagrin). Academics, we have frequently been told, have no grasp of the realities of the practice of safety reduction: they have a tendency to be overtheoretical, to value hypothetical models over concrete models, to look for abstract laws as opposed to dealing in the specifics of given situations, and so forth.

Assuming, for the sake of argument, that such criticisms are made and that they have at least a modicum of validity, it is worthwhile to spend a few pages discussing how this situation has arisen. This is useful not only in itself, but also because it will help to elucidate the difference between the standard approach to safety management and the new methodologies we felt we had to develop to do our jobs properly.

We begin at the beginning, then. Generally, we take it, most academic writers on safety management explicitly set out to make the study of safety more academically rigorous and therefore more scientific. But, what do we really mean by *scientific* in this context? In actuality, what do we really mean by *science*?

1.1 REFLECTIVE PRACTICE AND SAFETY PRACTICE

To answer these questions, let us (in the classic academic manner) avoid it for a moment and look instead at the assumptions that tend to animate discussions on rigor and science, especially as these relate to the soft (or social, or human) sciences. In this section, we draw heavily on ideas developed by Donald Schön in his classic work *The Reflective Practitioner* because it deals with many of the issues with which we dealt in our careers as safety (and psychology) practitioners. Schön argued that contemporary experts (he called them professionals) tended to use the jargon or rhetoric of what he termed *technical rationality* (TR) to justify what they do. In this view, *expertise* consists of "instrumental problem solving made rigorous by the application of scientific theory and technique" [1, p. 21]. This view is clearly derived from the widely held view that physics, chemistry, and (to a lesser extent) biology (the hard sciences) are the only true sciences, and that all other fields of knowledge should aspire to be as much like them as possible.

However, the belief system (or ideology) of TR presupposes a number of things. Moreover, these presuppositions are rarely stated openly, despite the fact that they have consequences for the way safety science (or any science) is carried out. These assumptions concern science: what it is and how it should be done. Specifically, TR presupposes that science is the disinterested discovery of objective laws of nature that are not context specific (i.e., are universal) and should be mathematically expressed if possible. This view has a number of corollaries; specifically, *all* knowledge should therefore be mathematized as much as possible, which means that quantitative data should be used rather than qualitative data and so forth. As we shall see, this view is deeply rooted in Western thought and can be traced at least as far back as Hobbes and Bacon (and possibly as far back as Plato). However, perhaps the deciding moment at which this viewpoint became the dominant ideology in the West was the publication of Isaac Newton's *Principia Mathematica*. This can be seen best by looking briefly at the "debate" between Newton and Goethe over the nature of light (the word *debate* is in quotation marks because, of course, Newton was long dead by the time Goethe criticized him; nevertheless, Goethe did face criticism from Newton's heirs, so the word seems appropriate).

Newtonian views "triumphed" in this debate, and importantly, it was not only Newton's view of the nature of light but also his view of the nature of science that prevailed. Newton's triumph was that he discovered objective mathematical laws that could be used for prediction. So, from the 18th century on, following Newton, science has generally been taken as the discovery of mathematical laws of nature that have an abstract, objective, and deterministic nature. It was not that experiments were not important, but the role of experiments was implicitly (and rather subtly) downgraded. In this view of science, what happened first was the discovery and (mathematical) elucidation of objective laws: there would then follow a (single) experiment that would prove whether the law was true or not. In this, Newton followed in a long tradition of Western philosophizing of science, going back to Plato but having its most aggressive proponent in Galileo, who had argued that "reality was best conceived of as consisting of separable parts in motion according to pre-established laws" [2].* Newton, to repeat, accepted the role of experimentation but only as a test that would prove once and for all whether the law that had been discovered was the one true law that fit the facts, but also (assuming the law was true), even more weakly, that it was merely a *demonstration* that the law was true.

---

* This view seems so fundamental to the materialist view of modern science that few see the hidden metaphysics behind it; in fact, Galileo explicitly took his philosophy from Plato. In the famous allegory of the cave, Plato argued that there was a real world of ideal forms behind reality. Therefore, mathematical abstractions (the discovery of which Galileo took to be the task of the scientist) were the true Platonic reality, and these consisted of atomized facts that were related to each other via (equally metaphysical) laws [3].

It was this view of science that Goethe rejected. As Ribe and Steinle put it:

> Newton's and Goethe's respective approaches … illustrate two very different approaches to experimental research. We call them theory-oriented [i.e., Newton's approach] and exploratory [i.e., Goethian] experimentation. Theory-oriented experimentation is often regarded as the only relevant kind: it corresponds roughly to the "standard" view in the philosophy of science that experiments are designed with previously formulated theories in mind and serve primarily to test or demonstrate them … By contrast, exploratory experimentation has been relatively neglected by historians and philosophers of science. Its defining characteristic is the systematic and extensive variation of experimental conditions to discover which of them influence or are necessary to the phenomena under study. The focus is less on the connection between isolated experiments and an overarching theory, and more on the links among related experiments. Exploratory experimentation aims to open up the full variety and complexity of a field, and simultaneously to develop new concepts and categories that allow a basic ordering of that multiplicity. Exploratory experimentation typically comes to the fore in situations in which no well-formed conceptual framework for the phenomena being investigated is yet available; instead, experiments and concepts co-develop, reinforcing or weakening each other in concert. [4, pp. 44–45]

Now (and this explains the relevance of this debate to safety science), Newtonian science works well when the systems under analysis are simple. But when they are complex and not easily reduced to simple mathematical models, then "[we] often start with a multitude of empirical findings whose interconnections and underlying principles are unclear. [We] must use experiments not so much to demonstrate propositions as to develop the concepts needed to make sense of multiplicity. The traditional isolated experiment is of little help here" [4, p. 45]. This is discussed further in Chapter 4: it is precisely because the impact of the single, isolated scientific experiment has been overemphasized in safety science that so much of it is, paradoxically, of dubious scientific value.

Goethian science is far less well known than Newtonian (that is, in the standard academic descriptions of "how science works"), and yet this exploratory approach, in which one experiment leads to another experiment with an aim to look at relations *between* the data elements studied (i.e., to look *at* what emerges from the data rather than attempting to look *through* the data to see the laws, causes, or cognitive architecture that lie behind them), is arguably more common in practice than the TR approach. Ribe and Steinle [4, p. 43] argued, for example, that Faraday, in his groundbreaking work on electric current, used an approach far closer to Goethe's than Newton's.

It should be made clear at this point that Newton's theory-first approach, in which experiments come second and then only to validate (or not) abstract mathematical models, follows from a classic division or dichotomy in Western thought, that between theory and practice. As Schön put it, this division, of abstract, theoretical (mathematical) knowledge and concrete (experimental, pragmatic) knowledge, with the former privileged, is even now reproduced throughout the Western education system, in which "'general principles' occupy the highest level (of prestige) and 'concrete problem solving' the lowest" [5]. This has led to a certain disdain among some physicists for the actual hard work of practical, experimental research. The classic example in the 20th century is the fame of Einstein. Newton prioritized the discovery of objective scientific laws as the basic aim of science, but Newton was also an experimentalist of genius. However, by the 20th century, Einstein could become famous as a physicist despite the fact that he never experimentally tested *any* of his hypotheses. The name of Eddington is less well known, although Einstein's theories, brilliant though they were, were merely hypotheses until tested by Eddington.*

This emphasis, we argue, has been damaging enough. However, perhaps even more dubious is the belief (seldom stated outright, but perhaps more widely believed than expressed) that science that does not attempt to discover "objective," "timeless" "mathematical" "laws" *is not really science.* As Crick put it: "Physics and chemistry are sciences. All the rest is social work" [7]. Few people would make this point so strongly. But, the emphasis on TR, with its insistence on abstract theorizing over concrete experimentation, provides a climate in which such an opinion can be expressed. As most people know, physics lends itself to approach in terms of abstract mathematical laws far better than biology, let alone sociology or the other social sciences.

To repeat, in theory (and how apt that word is in this context) the TR view of science accepts experimentation as an essential part of the scientific process. In practice, however, it downgrades it: TR assumes that the theoretical physicist comes up with the law, which is then tested experimentally. If the experiment proves it, then the law is accepted, but it is never forgotten who discovered the law in the first place. Einstein is famous; those who tested his laws are almost unknown. This view (and it is only one particular view of the scientific process) has implications for experimental practice, as we see in Chapter 4. But, it also has implications for other scientific methodologies. TR downgrades experimentation, but it completely *ignores* observational studies (on the grounds that these relate only with difficulty to abstract laws), implying that these are not really science.

But this is ridiculous. As McComas pointed out: "Copernicus and Kepler changed our view of the solar system using observational evidence derived from lengthy and detailed observations … but neither performed experiments." Moreover, "Charles Darwin punctuated his career with an investigatory regime more similar to qualitative techniques used in the social sciences than the experimental techniques commonly associated with natural sciences" [8]. The *Origin of Species* uses no mathematics, and as noted, Darwin did not carry out experiments. But, to claim that therefore Darwin was not really a scientist is absurd. Instead, Darwin's example demonstrates that TR only indicates one way of doing science, that there are others, and that experimental and nonexperimental (observational) studies may be as beneficial to science (or more so) than theorizing. It is also clear that once one accepts that observational studies (and similar methodologies) can be as scientific as any other

---

* Ironically, perhaps this is justified in that it has been argued that Eddington's work was flawed and inconclusive, and that Einstein's theories were not genuinely proved until the 1960s. This leads us to the social construction of science, which is discussed in Section 1.1.1. The key point, as shown later, is that Eddington's work, whatever status one gives it, was *observational*, not experimental. Observational studies have always been a key methodology in the hard sciences and not just the social sciences, despite what standard accounts would have us believe [6].

method, then the distinction between the hard and soft sciences begins to break down, and Crick's comment can be viewed as a highly misleading piece of rhetoric.

To clarify our position at this point, we are not specifically arguing for the creation of a Goethian safety science. What we are arguing is that there are *alternatives* to the textbook view of how science ought to function. We are also arguing that, before adopting a philosophy/methodology of science, the obvious question should be, Is this methodology the most appropriate for this subject matter? In our experience, it is almost invariably assumed within the field that TR is the only methodology available. This is false.*

1.1.1 THE SOCIOLOGY OF SCIENCE

What else is wrong with the view that "only TR = *real* science"? First, the development of the sociology of science, especially work by Thomas Kuhn, Bruno Latour, Donald Mackenzie, and others, has shown that scientists are not in fact the disembodied, disinterested automatons searching for facts presupposed by the ideology of TR.

What sociological studies of science have also shown is that science (like other behavior) is not so much a set of beliefs as a set of practices. Therefore "doing science" is not that much different from doing any other profession, like banking, teaching, or working in a DVD rental shop. In other words, there is a certain sociocultural environment in which practitioners work; there are political battles (jockeying for power and position); there are conflicts of class, gender, and ethnicity in terms of what gets said and who gets to speak; and so on. To misquote Karl Marx: "It is not 'science' which uses men as a means of achieving — as if it were an individual person — its own ends. Science is nothing but the activity of men in pursuit of *their* ends" [9; italics in original] (originally Marx was talking about history). *Science* does not create knowledge or tell us anything. People create knowledge, and not bloodless "rational" people either, but real flesh-and-blood people with hopes and fears and biases and a culture within which they function. Once this view is accepted (and there is a huge amount of empirical data to support it [10]), the view of science as primarily an inductive process (induction is the view that one should approach a situation with no theory but instead infer a theory from the facts) becomes highly problematic. Instead, it seems that most (perhaps all) scientific activity is about finding things out through trial error; we always have presuppositions, ideas, or imaginations about our subject before we begin our experiment, before we even begin to formulate theories (this was certainly the view of Karl Popper, among others) [11]. Needless to say,

---

* One other part of human life that is hard to classify in the TR view is simply practice, as in ordinary human practices. What, for example, are plumbing, working as an electrician, cooking, working a lathe (to choose among many possible examples) in this view? Either they must aspire to being as sciencelike as possible (e.g., rechristening cooking "domestic science") or they are simple random unstructured practices with no real methodology or coherent philosophy. Again, TR implicitly (therefore) denigrates the ordinary practices of ordinary working life. In a safety context, this has meant that working practices such as safety management have aspired to be as much like science (which, to repeat, means in this context theoretical physics) as possible, or else safety managers have simply been left to get on with it. Whether theoretical physics is a very good model for the complexities of safety management is a question that is neither asked nor answered in the TR view.

these presuppositions influence what is investigated, how we investigate things, and perhaps even how the findings of experiments are interpreted.

1.1.2 THE PHILOSOPHY OF SCIENCE

Second, there are the changes in the philosophy of science that have occurred since Thomas Kuhn wrote *The Structure of Scientific Explanations* (Cambridge University Press) in 1962. Karl Popper was the last major philosopher to attempt to create a benchmark law of how *all* the sciences ought to operate (his falsification thesis), and it is generally agreed that he failed, not least because his theories do not accurately describe how real scientists actually behave. Instead, perhaps we should see the sciences as a network of family resemblances. This idea was discussed by the philosopher Ludwig Wittgenstein, who wrote:

> Consider for example the proceedings that we call "games." ... What is common to them all? — Don't say: "There must be something common, or they would not be called 'games'" — but look and see whether there is anything common to all. — For if you look at them you will not see something that is common to all, but similarities, relationships, and a whole series of them at that. To repeat: don't think, but look! [12]

In the same way, *look* at the sciences. What is there *really* that links archaeology, physics, environmental science, and linguistics? We can only agree on a common "essence" (e.g., use of experimental prediction) if we decide that some of these (e.g., linguistics, astronomy, or archaeology) "aren't really science." But, by what criteria do we do that?*

Increasingly, philosophers and sociologists admit that there is no common essence. There is no single entity called *science*. Instead, there are only "the sciences," discursive and physical practices that people engage in at certain times and in certain places. And, in the last few decades philosophers have applied the same leveling tendency to mathematics and statistics [14]. Many contemporary philosophers now argue that these subjects also are discursive practices derived from embodied experiences. Like the sciences, logic and mathematics are best seen as activities in which people engage in a social context rather than a set of cognitive beliefs.

If this is the case, then the sciences are better seen as a web of practices rather than a pyramid with theoretical physics at the top and sociology somewhere near the bottom [15]. Practices suitable for one situation may not work in another. Solutions to problems are always context specific and should be generalized with care. Mathematics is not a panacea but a tool that should be employed when it is useful and avoided when it is not.**

---

* McComas pointed out the numerous studies that have demonstrated there is no essential scientific method. He also pointed out that the myth that there was such a method seems to have originated with the statistician Karl Pearson in the 1930s (although there were obvious precursors). Pearson was one of the founders of the modern frequentist statistics. We show in Chapters 3 and 4 that this is not a coincidence: frequentist statistics draw heavily on TR ideas [13].

** See Paul Ormerod's *The Death of Economics* [16], in which he argues that economics has been crippled for its search for mathematical models that are given precedence over empirical research.

Moreover, contemporary philosophers have also demonstrated that the fundamental axiom of the TR viewpoint (the fact/value distinction) is misconceived (TR presupposes that there are neutral objective facts that can be separated easily from subjective values: the task of science is to create predictive models of the former). The philosopher Hilary Putnam showed that these categories are hopelessly intertwined: almost all factual statements have a (more or less) hidden normative content (less intuitively, most evaluative statements also have a factual content) [17]. For example, is the use of the word *cruel* a value judgment or a fact? Putnam argued, following a long tradition in philosophy, that it is perfectly factual to describe Hitler as a cruel man even though the use of the word is profoundly entwined with its value-laden nature. Moreover, it can be argued that modern science would simply disintegrate if there were not a broad agreement on more or less ethical issues; that is, it is wrong to lie (misstate experimental findings), cheat (wrongly claim credit for others' ideas), and so on. Therefore, any inquiry into facts will also be an inquiry into values. More important, an inquiry into the facts of a matter will presuppose and be intertwined with not only the values of the scientists, but also the values of the culture in which the inquiry takes place. This of course is particularly apparent in safety science, for which values and scientific practices are deeply intertwined: it would be a strange safety manager who actually pledged to increase accidents or who argued that it did not matter how many people were killed as a result. A certain ethical approach is built in to safety science; it is not, and could never be, objective in this value-neutral sense.

Putnam drew an important conclusion from the disintegration of the fact–value dichotomy:

> There is a distinction to be drawn (and one that is *useful* in some contexts) between ethical judgements and other sorts of judgements [… (in the same way that)] there is a distinction to be drawn (and one that is useful in some contexts) between chemical judgements and judgements that do not belong to the field of chemistry. But nothing metaphysical follows from this fact/value distinction (in this modest sense). [18]

What Putnam is arguing here is that there is no objective distinction between facts and values.

Instead, it is a categorization, which is useful in some contexts and not in others. We may go further (before backtracking and returning to this subject later) and state that perhaps instead of a digital or binary phenomenon, the fact–value distinction is fundamentally an analog distinction that it is sometimes useful to treat as if it were binary. However, we leave this seemingly gnomic statement for the moment and return to it in the next chapter.

1.2 ABSTRACTION AND SAFETY SCIENCE

The reader may find it ironic that a book that began by arguing against the academic's tendency toward abstraction has immediately launched into what must seem to be a highly abstract discussion of the philosophy of science, but it had to be done if only to dispel the notion that science consists of the discovery of (timeless, mathematical) laws, and that if there do not seem to be any laws, then you are not really doing science. However, it still invites the question: what does this have to do with safety?

The answer is that, as we show, safety science has drawn most of its assumptions from TR (i.e., Newtonian science). But, as we argued, Newtonian science only works in highly specific circumstances, and it is by no means clear that safety science (or psychology in general for that matter) is one of these circumstances.

The specific problem here is that of unique versus common events. In an absolute sense, all events are unique. However, some can be considered, to all intents and purposes, as having a common essence that permits prediction. For example, we can predict how fast a ball dropped from a certain height will fall and how hard it will hit the ground. We cannot do it with complete precision, but the imprecision or fuzziness is not fatal to the prediction (in most contexts); if we repeat the experiment with a different ball, the results will probably be more or less the same. The fact that scientific experiments (and the laws that explain them) actually function by the principle of *ceteris paribus* (i.e., all other things being equal) is sometimes glossed over in histories of science, but it is self-evidently true. Newtonian physics simply predicts how objects would behave in an abstract mathematical world. Therefore, the results that we get from real experiments are never exactly the same as those predicted because of external factors (e.g., in terms of the laws of motion, things such as air pressure, temperature, and so on affect the experimental [observed] result). But for purely pragmatic reasons, this does not really matter for most physicists because in simple experiments the external factors (context) can be controlled, and the experimental results for similar experiments tend to be similar.*

However, in terms of predicting behavior of animals or humans (let alone the complex sociotechnical breakdowns that we term *accidents*), not only does this fuzziness increase, but also it starts to matter. In absolute terms, of course, the same law of *ceteris paribus* (that laws only work if the context can be controlled) applies to the laws of human behavior, and in that sense they are no different from the laws of physics. But, the various contexts in which human behaviors take place seem to have even more effect on the efficacy of predictive psychological laws than do the equivalent contextual effects on the behavior of (for example) celestial objects. There is no mystery regarding why this is the case: humans (like all living things) are active, not passive. They actively transform their environment, as opposed to passively *reacting* to it, and their behavior is in turn transformed by the environment [20]. The relationship between a human and his or her environment is therefore dialectical rather than merely easily encompassed by the one-way metaphor of cause and effect.

This is not to deny the possibility of psychology/sociology or safety science as a genuine science. Human behavior can be predicted (if not in a deterministic sense, at least in a statistical sense; see below). But, as with the laws of physics (except even more so), the laws of psychology/safety science are always and will always be affected by context. In other words, we can make predictions when we know enough

---

* Or, if context cannot be controlled, we simply give up. Cartwright [19] described how even the most basic laws of physics fail to predict simple events. For example, take a dollar bill, go to the top of a tall building, and drop it. Where will it land? There is no physicist who has ever been born (nor will there ever be) who can tell you. Laws are only predictive when we can "bring the lab out into the world" as much as possible.

about the context in which the event or behavior is going to take place. The laws of safety science/psychology are context specific, not universal.

1.2.1 LAWS AND REGRESS

If we accept this, we see a problem for TR that depends on the assumption of timeless, context-free, deterministic laws. As well as problems with the context-free assumption of the TR equation, there are philosophical problems with positing *timeless* laws that universally explain phenomena per se, for example, the problem of infinite regress. If timeless laws are how things are to be explained, then presumably the laws of physics must themselves have been created by timeless laws, but this then means that these other laws must also have been created by laws and so on. As a result of these criticisms, some philosophers have championed the "framework" conception of the laws of the sciences. Under this conception:

> Laws provide a framework for building models, schematizing experiments, and representing phenomena. Laws, moreover, have very broad, but *not* universal, domains of application. Rather than taking laws to be universally true and delimiting all possible worlds, the proponent of the framework conception takes them to be broadly *reliable* for a wide array of tasks. [21; italics added]

The reader who has followed us this far will doubtless guess that this view of laws as heuristics or guides to action is more compatible with the view of the sciences for which we argue than that which sees them as timeless and objective. But, the framework view (sometimes termed *instrumentalism*) clashes with certain views of science that originate, it may be argued, from the very first arguments about Western metaphysics: those between Plato and Aristotle.

There is no room to go into this debate in any great detail. However, it is sufficient to say that the debate between Plato and Aristotle is not just a debate between rival views of philosophy but between rival views of science. Plato posited the existence of Universals, timeless metaphysical truths that exist outside and beyond time and space. Euclid also argued for the existence of such laws and argued that mathematical laws were laws of this sort [22]. The true task of the philosopher was to search for and describe these Universals. Moreover, as the famous allegory of the cave illustrated, Plato saw the world of these laws as the real world. Therefore, the true philosopher or scientist was disinterested, a speculator not a doer, and worked out mathematical and philosophical laws from first principles; as we saw briefly, Galileo and (implicitly) Newton also drew on this tradition when they created their view of what sciences ought to be.

It would be a gross oversimplification to see Plato as the patron saint of the disinterested intellectual as opposed to Aristotle as the patron saint of people who wish to get their hands dirty. And, there are many aspects of Aristotelian philosophy with which we would disagree. Nevertheless, there is something in the comparison. Whereas Plato emphasized disinterested speculation, Aristotle made observed reality the basis of his philosophy: his observational studies lie at the beginning of the Western tradition of biology and physics. In addition to this, Aristotle argued that Plato's timeless Universals were in fact immanent in perceived empirical reality;

that is, they are simply inferences we make from the real world in which we live and act every day. This view was developed in the Middle Ages to create the philosophy of nominalism, which stated that the so-called Universals were merely linguistic artifacts. Therefore, for nominalists, only individual objects really exist; the relations between them (whether in the form of laws or of Universals) do not.

It is far too simplistic to draw a simple binary opposition between Platonic and Aristotelian conceptions of science. Aristotle took over much of his philosophy from Plato, although he adapted it in the process. Nevertheless, as long as this dichotomy is not taken too literally, it will hopefully help explain the concepts discussed in terms of Newton and Goethe. In his philosophy, Plato stressed mathematization (a view he probably took over from Euclid), speculation, and timelessness. Aristotle stressed observational studies, classification, and actually getting out into the world to do some research. It is clear that TR derives from the Platonic, not the Aristotelian, view of science.*

There is another Aristotelian concept that it is useful to discuss briefly here: *phronesis*. Phronesis is the word used in Nicomachean ethics to describe what Aristotle called practical wisdom, and that we might define in our own age as common sense. Phronesis comes from experience, but it is not formed by abstracting laws or regularities from reality. Instead, it is formed by having access to a library of *cases*, in other words, by having seen and done many things that are relevant to the situation at hand (see Chapter 8 for a further discussion of this). But, for Aristotle, it is also tied in with character and having a moral sense: there is no question of being detached or objective in a value-neutral sense (in other words, it does not presuppose the fact–value distinction). We might argue that professionals such as doctors, nurses, or even plumbers and electricians make use of phronesis. The key point is that phronesis is in the world, not out of it; it engages in practical actions, not creating theoretical models [23]. We return to this concept in the final chapter when we discuss our definition of *expertise*.

The next question the reader will perhaps still be asking is: so what? So what if there are Aristotelian and Platonic views of science? The relevance of understanding the difference between these two philosophies, we argue, is that when safety scientists argue that their subject should become more scientific, it tends to be the Platonic/Newtonian conception of science they have in mind. Thus, when one comes across statements (as happens rather often) such as "Human factors must become more scientific" or "Safety management is not yet a science but may well become so," then what the authors have in mind is that human factors/safety management must work harder to develop timeless Platonic laws that are universally applicable.

As the reader will no doubt have guessed, we would question this approach. First, it is not at all clear that even the laws of physics or the practices of physicists

---

* In case any Aristotle specialists are currently wincing at this point, we are well aware that there are huge aspects of Aristotle's philosophy (e.g., his notion of essences) that are more "Platonic" in terms of the schema we have built up than we are letting on. That is true, but we are merely making these divisions for simplicity's sake, and though not the whole truth, they contain an element of truth. It was, after all, Aristotle, not Plato, who actually went out and performed the first biological and physical observational studies, generally agreed to be the first serious scientific endeavor.

are accurately described by the Platonic ideal. More specifically, we have the question: Are accidents and safety issues in particular amenable to this form of analysis? A safety researcher has a number of problems that the physicist (or even a biologist) does not have. To illustrate these, we turn to the realities of the second of the main tools of the scientist: not theory, but experimentation.

1.2.1.1 Problems with Experimentation

Remember the distinction we drew between one-off events and essentially repeatable events. It is (comparatively) easy to model simple, repeatable events (such as a ball falling from a table) and much harder to model things when humans are involved. But, when we are dealing with complex, sociotechnical failures (e.g., *Challenger*, Three Mile Island, and the Ladbroke Grove rail accident in 1999) that involve both humans and fantastically complex technical issues, the problems become harder still. The key problem is that these events are to a certain extent one-offs. We cannot turn back the clock and rerun them to test our hypotheses regarding why they really happened. In other words, due to their complexity, they are essentially unrepeatable.

This is made worse by the phenomenon of synergy (often mentioned in business studies but rarely discussed in safety management theory). Synergy occurs when elements combine to create emergent properties that are not present in the individual elements. For example, when we combine blue and yellow paint to create green, there are no green elements in the blue or yellow (and we cannot split the green into its component blue and yellow parts again to study how it came about). Green is an *emergent property* that occurs when the two colors are mixed (needless to say, there may be emergent or synergistic properties of more than two elements).

In the same way, causal elements that are safe on their own may become risk factors when (and only when) combined with others; that is, their effects are synergistic (we discuss the concept of emergence in Chapter 6). Certainly, some safety theorists touch on this issue when discussing latent pathogens and proximal triggers. But, there need not only be one latent pathogen or, for that matter, proximal trigger. There could in theory be many. In this case, how does one determine the extent the phenomenon was or was not a causal factor? How can one know that the accident would not have happened if the element or factor had not been there? Remember, we cannot turn back time and remove individual elements in an experimental or quasi-experimental setting to see which was the primary cause or decisive (causal) factor. It is precisely this form of experimental validation that physics and chemistry rely on to validate their own laws. Experimental validation is not something that can simply be dispensed with. It is an integral part of the process. The proof of a physical law is precisely in that sense: we can rerun experiments and remove causal factors to see which the decisive one is.

We cannot do that in safety science. So, how then can we ever prove any of the laws of safety science, regardless of whether they are true or not in some abstract philosophical sense? The nature of a scientific law (e.g., in physics) is predictive. Regardless of whether we view the laws of physics as metaphysical (i.e., timeless and objective) or man made, we can still use them to predict with a fair amount of accuracy (for example) where the planets in our solar system will be in two years time. In the same way, in the case of safety science, we would like to create predictive laws: in other words, we would like to say that if we bring together the causal factors x, y, and z, an accident will inevitably result.

With gigantic and massively complex events (e.g., the *Challenger* disaster), given the inherent unrepeatability of the event, we could never know this (that is, we could never prove it in any genuinely scientific use of the word *proof*) because we do not just have x, y, and z but an almost infinitely large number of causal factors. At best, our law would remain merely a hypothesis. At worst, it would be positively misleading, creating pseudocertainty.

Now it has to be admitted that few people currently deny that such events are difficult to predict. Instead, it is argued that specific subfeatures of the event (e.g., the action of a particular valve) can be studied experimentally and therefore predicted. There are a number of problems with this argument as well.

1.2.1.2 Ecological Validity

First, there is the problem of ecological validity. This is often sidestepped in discussions of experimental methodologies. It is, however, crucial. This relates to the degree to which an experiment successfully models the real world, such as whether the behavior of a human in a laboratory will accurately capture the behavior of that same person in a real situation. Again, as with the examples above, technically this applies to all aspects of science. For example, in psychology, one could argue that the behavior of a participant performing a train simulator task is not at all similar to the behavior of a train driver in a real train. Certainly, with the development of more and more advanced simulators, this problem can be dealt with to a certain extent, but in the final analysis, the participant in the simulator will always know that the simulated situation is not real, whereas the train driver will always be aware that his or her situation is only too real. This is particularly apposite to studies of stress, aggression, and fear, the key emotions in terms of accidents.

This is not an abstract point. As Diane Vaughan made clear in her classic study of the space shuttle accident, *The* Challenger *Launch Decision* (to which we return in Chapter 7), even in the context of engineering, engineers are well aware that an experiment in a lab on a piece of machinery is not necessarily going to be an accurate predictor of how that piece of machinery will work in the very different context of outer space [24]. An act of inference and interpretation must always take place, and it is by no means certain that this interpretative act will end in consensus. In the case of *Challenger*, it was precisely the interpretation of experiments concerning the o-rings that led to the disaster, with some engineers arguing that the experiments showed they were safe, and others arguing the opposite. Ultimately, it was the people who managed to gain a social consensus on the interpretation of the experiments who won the argument. (This has been confused in that it is now obvious after the fact that safety was compromised. At the time, this was not nearly so clear: the data were ambiguous because, in fact, all experimental data are inherently ambiguous in that they must be interpreted before they can be understood.)

1.2.1.3 Hermeneutics

This emphasis on interpretation brings us to hermeneutics, which we have discussed extensively elsewhere [25]. *Hermeneutics* is the science or study of interpretation, and in this view, it is axiomatic that all texts are open to multiple interpretations (in other words, to use the jargon, they are polysemous). It is less often pointed out that the laws of hermeneutics apply not only to texts, but also to behaviors and events, and that this includes experimental data. Experiments do not speak to us by passing truth directly into our minds. Instead, they are interpreted by humans in a sociopolitical context. More crucially, different people will interpret them in different ways at different times. Again, this is something, doubtless, that is true of all scientific endeavors, but it is particularly true of safety science because this is a particularly fraught area of inquiry, with considerations of politics and money at the fore. A safety scientist who attempts to make a prediction (as, we are taught, scientists should) will quickly be in hot water if the prediction is not of the right sort and may be overruled for anything but scientific reasons (again, the *Challenger* disaster is the classic example of this). This does not even touch on other issues relating to science that again are particularly relevant to safety management (i.e., difficulties with genuine replication of experiments, as well as issues of funding [specifically as relates to what experiments are carried out] and more general cultural biases).

Now, all this may appear self-evident, and the key point, the impossibility of *replicating* complex disasters such as those at Three Mile Island or Chernobyl in the laboratory (or anywhere else), is well known. Nevertheless, the implications of these phenomena (that it is not at all clear that laws can ever be created that will predict such events with any degree of reliability, and that even if they could, they could not be proved by experiment) are less often examined, and when they are, the discussion frequently concludes with the pious wish that, although such laws do not exist at present, they may exist at some point in the future (and safety science will then become a proper science). But, is this necessarily the case? Do we actually want such laws? What use will they be? Will they actually bring any practical benefit?

On the other hand, if we decide that such laws do not exist, whither safety science/management? Do we simply give up? Or, do we, on the contrary, explore different methodologies that, although they adopt more of a getting-your-hands-dirty, Aristotelian approach, may nevertheless produce concrete results?

Now, it might of course be interjected at this point that we have set up a straw man because few, if any, contemporary researchers openly search for the timeless laws of safety science. In fact, has this viewpoint not been abandoned as hopelessly naïve and beside the point?

1.2.2 MODELS

We would argue that, even when this search has allegedly been given up, it is smuggled in again via the back door. For example, many classic texts of safety management (at least in an academic vein) deal with issues of models (usually models of accidents or other unwanted events). However, it is noticeable that most of them provide a single (generic) accident model, and it may be argued that the relations between the component parts of these models tend to have lawlike regularity. In other words, the search for one model to model all accidents implies that such a model might exist, that it might accurately represent the truth about some or all accidents, and that this model therefore can function as a de facto law, a context-free representation of the essence of an accident.

However, it is not clear that all (or even most) accidents can be represented by just one model. Further in this chapter, we look at accident models and even argue that some are better than others. However, this invites the question: is it even possible that one model will successfully apply to all accidents? The answer may be "yes," but we have never seen the argument in the literature that it might not be. It is simply assumed by those using the rhetoric of TR that the answer will be yes. Why? How do we know this is not a family resemblance situation for which there are many different phenomena that are merely classified or categorized as accidents for primarily sociological or psychological reasons?

This is confused in the literature by the situations that are used as examples. Referral tends to be to the same accidents: *Challenger*, Ladbroke Grove, Three Mile Island (an accident so well known it has its own abbreviation, TMI). In other words, referral is to major breakdowns in high-consequence, high-technology industries. However, who is to say that these are the only kinds of accidents? For example, the new science (or study) of famine studies has shown that, far from existence as natural events, famines are in fact breakdowns in socioecological systems, and that an analysis of systems, culture, ecology, and politics is necessary to explain fully why they happen [26]. In the same way, epidemiological disasters (such as the spread of the human immunodeficiency virus) are rarely discussed as accidents, although they also are breakdowns in complex systems with human input. Inclusion of these phenomena as well as the more familiar disasters would question the idea that a generic accident model will fit all cases.

Moreover, as Schön wrote: "Technical Rationality depends on agreement about ends. When ends are fixed and clear, then the decision to act can present itself as an instrumental problem. But when ends are confused and conflicting, there is as yet no 'problem' to solve" [27]. What is the aim of safety science? Is it to reduce accidents to zero? Presumably not. That, it is assumed, is impossible. Therefore, the aim must be to reduce them to an acceptable level. But, who decides what is acceptable? Should cost be taken into account when calculating this level? How do we know when we have reached it? When we have reached it, do we stop attempting to reduce accidents any further, or do we see if there is still room for improvement? How quickly should we attempt to reduce accidents? What methodologies should be used? Should we, for example, like H. W. Heinrich [see 50] (the godfather of the field), attempt to empower management as much as possible and assume that if managers have complete control over workers, then we should see a reduction in the accident rate? Or, should we empower the workers? Should we punish human error (as Heinrich argued)? Or, should we (as Gerald Wilde [28] argued from within the school of risk homeostasis) reward the lack of error? These questions demonstrate Putnam's [see 17] point: in the real world, facts and values are intertwined, rarely more so than in discussions of safety. *TR presupposes that facts and values can be separated.*

When there is fundamental disagreement about means and ends, Schön argued that TR (what we called Platonic or Newtonian science) breaks down because this implies a discussion about values, politics, and morality, and these are topics technical rationalists are particularly ill equipped to discuss (remember that TR deals only with facts, not with values).

The key point is the rigor-or-relevance debate. Academics and safety specialists can create extremely rigorous mathematical models of discourse, action, and safety behavior, but these have little relevance to behavior in the field and have little measurable impact on the real-world safety situation. Real interactions with real people, on the other hand, can have such an impact (as we shall see), but these interventions (frequently done in an ad hoc manner) tend to have little of the mathematical rigor we have come to expect from Platonic/Newtonian/technically rational science.

1.2.2.1   An Example of the Complexity of Causes

Perhaps an example will make these points clearer. There is an explosion in a city. Now, was this an accident, or an act of terrorism, or what? The answer is that at this stage we simply do not know; we do not have enough information to classify it. However, the "camera pulls back," and we now see that this city is Sarajevo, during the war in the Balkans (in the 1990s). Moreover, we now see that the building in question was hit by an American missile. Obviously, it is not really an accident at all: this is war, and such things happen. Then, we find out more. The building in question was the Chinese embassy, and it seems that it actually was hit by mistake. So, it turns out, this was an accident after all. But, then the wheel turns again. Some people argue that due to China's reporting of the war and its hostility to NATO actions, the Chinese embassy really was a target after all. So, it seems, if this claim is accepted, that the event *was not* an accident. As we might imagine, these claims are furiously denied.

So, was this event an accident or not? People wiser than us may know the answer to this question, but as far as we are concerned, all we can say is that the debate rages on. In other words, we may *never know* for sure whether this was an accident at all.

This example raises a few issues that are important. First, not only the cause but also the actual classification of an accident is context specific. We may see an event as an accident or not depending on our view of it, specifically, our view of the context in which it occurred. Change the context, you change the classification (or definition). Accidents do not preexist before they are classified as such — it is simply the case that some events have this label attached, and some do not.

Second, the attempt by academics to separate accidents from the sociopolitical circumstances in which they occur is futile, and this should be interpreted in a very strong sense. Judith Green has argued that the classification/categorization of an event as an accident is socially constructed [29], and it is hoped the above example will lend support to that argument. Whether an event is seen as an accident will depend not only on the context and how the context is interpreted, but also on the sociopolitical context of the event and of the observer. Marxists will interpret this event in a different way from NATO generals and Serbs in a different way from

Bosnian Muslims, and their classifications of the event will lead to different investigatory styles and therefore different causal factors and intervention strategies. If this is accepted, it becomes clear that the search for any one model (or law) of an accident that will cover every event that we classify as an accident is futile.

1.3 CAUSALITY AND ACCIDENTS

One of the key issues in terms of accident analysis is the classification *cause*. Again, this follows from the assumption of TR, for which our modus operandi is to break down any given object into its (discrete) component parts and then classify these components according to the causal laws that govern them (this is our friend reductionism). Again, we must remember that this classification (cause) always occurs in a certain context, and that we are always going to have certain prejudices or presuppositions as to what constitutes a cause depending on our background and the current situation. As the philosopher R. G. Collingwood put it:

> A car skids while cornering at a certain point, strikes the kerb and turns turtle. From the car-driver's point of view the cause of the accident was cornering too fast, and the lesson is that one must drive more carefully. From the county surveyor's point of view the cause was a defect in the surface or camber of the road, and the lesson is that greater care must be taken to make roads skid-proof. From the motor-manufacturer's point of view the cause was defective design in the car, and the lesson is that one must place the centre of gravity lower. [30]

Collingwood is arguing that the cause of the accident depends in this case on the socioeconomic background and career choice of the person making the attribution of causality.

However, despite these and other arguments about the social biases of attributions, many academics working in the field of safety still tend to argue (or at least tacitly accept) that the word *cause* is a reasonably unproblematic description of objective fact (it should be noted that many practitioners in the field do not agree and argue that cause should be used with extreme caution, if it is used at all [31]). Moreover, in the area of root cause analysis, a variety of methodologies exist that claim to help people discover the root cause of an accident. So, who is right? The academics or the practitioners?

1.3.1 PHILOSOPHY AND CAUSALITY

It was one of the key contentions of David Hume that we do not actually observe causality. We may observe an object coming into contact with another object or a person engaged in an activity prior to an unwanted event, but we can never actually perceive a causal effect directly. Instead, a union in the imagination is formed between certain events, and this (psychological) union is what we refer to as cause [32]. This is one of the (very) few arguments in Western philosophy that has not yet been refuted. The only conceivable response is to claim, as Kant did, that causality is a necessary prerequisite for understanding anything. Even if this is true (we argue below it is actually false), it admits the basic point: the causal interpretation of events is something we bring to a situation, not something that we infer from it because it is necessarily real. It is an active act of interpretation and therefore of classification.

1.3.2 THE STATISTICAL REVOLUTION

In a similar way, since the mid-19th century, the so-called hard sciences have largely abandoned deterministic (Newtonian) causality in favor of the nondeterministic statistical viewpoint. This shift is most usually discussed in the context of quantum mechanics, which demonstrated that instead of a particle being in a certain place, there is a probability distribution of where each particle might be found. Therefore, the world is probabilistic, not deterministic. This of course backs up Hume's point. As Bertrand Russell put it: "So far as the physical sciences are concerned, Hume is *wholly* in the right; such propositions as A causes B are *never* to be accepted" (first emphasis in original; second emphasis added) (it follows from this that statements such as "X caused the accident" should also never be accepted) [33].

However, despite that it is well known that causality has been abandoned in particle physics, the adoption of probabilistic descriptions in the rest of the hard sciences is less often the subject of remarks. Most scientists would now argue that deterministic explanations have only a small and local applicability and even then only with application of the *ceteris paribus* rule (as we discussed in Section 1.2) [34]. Instead, reality is probabilistic. Unfortunately, this statistical/probabilistic revolution has lacked an Einstein or a Newton to publicize it, so the repercussions of the statistical view in general are still not widely understood (to the extent that quantum reality is seen as the freakish exception to the general applicability of deterministic laws as opposed to the other way round). However, this should not obscure the importance of the revolution. As Hacking stated: "The erosion of determinism constitutes one of the most revolutionary changes in the history of the human mind" [35].

The key point of this revolution is easily stated: instead of a deterministic view in which certain causes inevitably lead to certain effects, in the probabilistic view certain causes may or may not lead to effects with differing degrees of *probability*. Therefore, contemporary scientists talk about statistical correlations as opposed to deterministic causality. This can be confusing because scientists still sometimes use the word *cause*, but when they use the word, it does not have the same meaning as it has in ordinary language. As Surry (someone who actually works in the field of safety rather than someone who theorizes about it) put it:

> The term [cause] is still used among scientists but … to merely indicate that there is an association between two events. This association is not necessarily linked by inevitable ties, and is not necessarily always true. Simply the association has been observed frequently. The problem arises that the lay usage of the term still clings to the older meaning: thus scientifically observed statistical causal relationships are misinterpreted by society. [36]

It is in speaking of this older, Newtonian view of causality that Pattee wrote: "It is not obvious that the concept of causation, in any of its many forms, has *ever* played a necessary role in the discovery of the laws of nature. Causation has a tortuous philosophical literature with no consensus in sight, and *modern physics has little interest in the concept*" (italics added) [37].

It is a failure to understand this fact, and to understand the nature of the laws of physics, that has helped lead to the misleading distinction between the hard and soft sciences. Unfortunately, as we have seen, some people even today continue to argue that the social sciences are inferior to physics or chemistry because the laws of sociology are not universal but situation specific, because they are not deterministic but probabilistic, and because sociologists (for example) tend to use observational studies, not experiments. But, this is based on a misconception of the nature of the modern sciences. In practice, the laws of physics also are situation specific (according to the framework view of these laws, discussed earlier). When physicists talk about causes they actually refer to the same statistical associations as used in sociology or psychology. Observational studies are not in fact specific to sociology or psychology. On the contrary, they form the backbone of medicine, geology, biology, and many other sciences. Observational studies are in no way an unscientific methodology.

1.3.3   ATTRIBUTION THEORY

We have questioned the standard (TR) view of science (one in which objective scientists discover [inductively] objective, context-free scientific laws). We prefer to view the sciences as a series of practices and have looked at a number of alternative ways of conceptualizing the sciences that would, perhaps, be better suited to safety research. We have then shown that modern science has little use for the concept of causality (as that word is generally used), relying instead on statistical association.

Following from this last point, given that the universe seems to be stochastic (probabilistic), it is clear that the attributions of causality that people make are more likely to be socially constructed than representations of reality. This applies equally to causes established by technocratic experts. If the findings of quantum mechanics and other fields are correct (and there is much evidence that they are), it is clear that any attribution of a deterministic cause by anyone will be, to a certain extent, a social construction.

However, now we can go further than that because psychologists have, since Heider wrote *The Psychology of* Interpersonal Relations (John Wiley & Sons, New York) in 1958, demonstrated biases in the attribution of causes. There are two that are of particular interest in the attribution of accident causation: the principle of similarity and internal versus external attributions.

1.3.3.1   The Principle of Similarity

First, there is the principle of similarity. McCauley and Jacques [38] noted that it has been argued that people tend to view causes as similar to events. On the most basic level, this means that if the event is big (i.e., has a large impact), then people will tend to look for big causes; if the event is perceived as complex, then people will tend to look for a large number of causes. The corollary of this is that if the event seems to be small, then we will look for small causes. The very logistics of accident investigation demonstrate this bias and show that it will tend to be the case.

For example, after major incidents (such as the Ladbroke Grove train crash or the *Challenger* disaster), major accident investigations take place. Dozens of witnesses are called. Millions of pounds or dollars are spent on reconstructing events, conducting trials, calling expert witnesses, and so on. This creates an environment in which a certain kind of answer comes to be expected. For example, what would the reaction have been if, at the end of the Ladbroke Grove inquiry, the multimillion-pound report merely stated, "The driver drove through a red light," and nothing else? The public outcry can be imagined. Instead, it was expected that systems failures would be found, and they duly were. In the same way, an accident in which someone falls and breaks a finger will tend to be classed as having a simple, direct cause although, conceptually, one could argue that such minor events always occur within a systems context as well.

So, we tend to look for big, sociological, or systems explanations for major accidents and small, individualistic, and proximal (frontline) explanations for minor accidents. This is why human error is an oft-cited cause for routine failures. This is an *investigation bias*, which has no necessary relationship to what actually happened.*

1.3.3.2   Internal versus External Attributions

Perhaps the most important aspect of attribution theory is the demonstration of sociocultural biases in attributions (i.e., attributions of causality) for events. For example, people in the West are far more likely to stress personal internal factors (e.g., dispositions, motivations, intentions) as causes than non-Westerners. To demonstrate this, Miller conducted a study in which Indian and American participants were asked to think of good and bad actions performed by various people and then to provide explanations for why they did them. The Americans were far more likely to provide internal explanations for events than the Indians [40]. This study has been supported by others studying, for example, the differences between Japanese and American attributional styles [41]. These biases go very deep. Lillard described how cultural differences go beyond particular aspects of causal chains to the very heart of the matter: the social construction of what actually constitutes a cause at all [42]. For example, Strauss described how Native Americans tend to stress environmental aspects for human behavior (i.e., see it contextualized in dynamic interrelationships with other people and situations) rather than focus on internal cognitive factors. Westerners (such as social workers and especially psychologists) find this difficult to deal with and sometimes conclude that Native Americans make excuses for their own behavior. But, there is no value-free position to take on this. Native Americans' interpretations are as objective as any alternative interpretations despite the fact others may not agree with them [43].

---

* For example, it is one of the axioms of chaos theory that small causes can lead to large events (e.g., the butterfly flapping its wings in Africa leading to a huge hurricane in North America). Less intuitively, large causes can lead to small events. For example, Fischer has shown that, despite what is generally thought, the defeat of the Spanish Armada was of no great consequence (historians, of course, feeling that such a great event *should* have a major impact, have long looked for events to link it with). Similarly, large systems failures in organizations do not *necessarily* lead to large accidents. They do not necessarily lead to anything [39].

There are two questions that follow from this. First, does this cross-cultural influence on causation apply to accidents? The answer would seem to be that it does. For example, Salminen and Seth discovered in a study of occupational accidents that Finnish workers and foremen were more likely to give internal explanations for accidents (e.g., to state that the victim had violated safety instructions or made a mistake) than Ghanaians [44]. As a general rule (not a law), non-Westerners are more likely to offer external attributions for events.

The second question is perhaps more controversial. We must ask: are scientists and experts free from attributional biases? The firm answer here is no.

First, there is an ontological reason for this answer: given nondeterminism in contemporary science, by what criteria could one decide if something constitutes a real cause or not? It is certainly true that scientists produce causal statements, but as we have explained, they are not using the word *cause* in its original sense (i.e., in a deterministic sense). It is even true that some people (usually academics) do use cause in a deterministic sense (especially in accident investigation), but here again we come up against the fundamental issue: given that accidents cannot be repeated exactly, how can one tell whether their speculations are correct or not? (This could only be proved by repeating the accident with the causal factor removed, and of course we cannot do that.)

There is also a considerable amount of evidence that scientists and experts display cultural and social biases in their explanations just like everybody else [45]. For example, Guimond et al. found that students in the social sciences were more likely to give external attributions for social phenomena (e.g., poverty) than engineers, and that this tendency increased throughout their courses [46]. Even the basic axioms of Western psychology (the idea that actions are caused by internal mental states) seem not to be universal but culturally constructed [47]. Increasingly, the very basis of modern cognitivist psychology is viewed in terms of the Western bias in favor of internal (personal or dispositional) states as causes (see Chapter 5).

More specifically, we can state unequivocally that these biases apply to safety experts as well. For example, Kathryn Woodcock and Alison Smiley carried out a study in which 16 safety specialists (with responsibility for accident investigation) filled in questionnaires to ascertain various psychological and demographic data and were then asked to carry out accident investigation simulations. The purpose of the study was to see if the investigative results obtained could be predicted from features of the investigators.

Woodcock and Smiley discovered that more systems factors were discovered during the investigation than human error (40% versus 25%). However, many of the systems factors identified were failures of personal protective equipment, factors that "have a subtle organizational utility, giving the appearance of impartiality/willingness to blame management without exposing management to severe liability" [48]. When personal protective equipment causes are allowed for, 37% of accidents were identified as human error, and only 27% were identified as caused by systems or organizational factors.

The two key predictors of willingness to provide worker-centric human error-type attributions were powerlessness and job position. The more powerless the experts stated they felt, the more likely they were to blame staff for incidents. In a similar way, the more senior the position of the safety specialist, the more likely the specialist was to provide human error-type attributions (powerlessness is a subjective feeling, job position an objective one, so there is no conflict here). Finally, subjects whose job involved capping injury compensation were more likely to produce human error attributions.

It is important to note that that the causes assigned by experts could be predicted given some knowledge of the people involved. This means the causes are not objective (i.e., independent of who found them). Stating that an incident was caused by human error (or anything else) can never be a motiveless, context-free, scientific description of truth but is always a motivated and functional judgment produced by individuals operating within a specific socioeconomic context. Put simply, if it is in the safety specialist's interests to produce a human error explanation, then the specialist will tend to do so.

Now, a common mistake when discussing attributional bias is to view it as *error*. Within this view, we are looking for the most objective account in which bias has somehow been ironed out. But, we would argue that notions of attributional error are now outdated. The quest for objectivity in this sense is futile. All accounts (causal or otherwise) are biased and must be contextualized. There is no such thing as objectivity here if objectivity is defined as a judgment made without presuppositions by someone with no motives or goals. We always have motives (which will influence not only what we look for, but also how we interpret what we find), and attributions are always produced in a context that will influence the attributions made. As Roger James put it (discussing the philosophy of Sir Karl Popper), it is frequently assumed "that truth can be discovered only by eliminating bias and prejudice. Popper shows this is *impossible*. Everybody is biased, prejudiced and interested (in both senses of the word)" [49]. Popper argued that it is not through an unbiased (lone) expert that truth is discovered, but through open and public discussion. We look later at this concept that truth is found via open and free debate between individuals (plural), leading to consensus. However, for the moment, let us return to safety science.

1.4 HEINRICH

To bring all this abstract discussion of the philosophy of science and safety back to Earth, we now look at the origins of the field, beginning with the man who, more than any other, can be credited with making, or attempting to make, the study of safety into a science: W. H. Heinrich.

It is perhaps ironic that the modern, scientific study of accidents begins with Heinrich as he was not a scientist but worked in the insurance industry in the United States in the 1920s. Nevertheless, in his pioneering work, *Industrial Accident Prevention* (originally published in 1931), he set out his own view of the methodology and philosophy of the fledgling science [50]. What may be surprising is that, in many ways, safety science has always followed (and continues to follow) in Heinrich's footsteps. It is hoped that whether this is for the best will be clearer by the end of this chapter.

At the beginning of his book, Heinrich stated his commitment to what we have termed the Platonic (or TR) view of science: "In all … science … one should expect to find governing laws, rules, theorems … these exist. They are stated, described, illustrated. After understanding comes application" [51]. In other words, science is the creation of abstract laws, and only after these laws have been discovered is the practical side of the study to begin. Moreover, these laws are "based on knowledge of cause, effect, and remedy … a scientific approach" [52]. Neither the linkage of the abstract law theory of science and deterministic cause and effect nor the idea that these things are axiomatic could be made clearer; we are simply supposed to accept them before we begin any observational or exploratory work.

Heinrich's methodology was to study the various accident reports available to him (filed by supervisors in various insurance firms with which he had worked) with a view to discovering the laws of accident causation, which could then be the basis for the laws of accident prevention.

However, he actually began by deducing some preliminary axioms he saw as self-evident and therefore felt no need to provide any evidence for them. He wrote:

> Whatever the situation or circumstance may be, it is still true, and as far as may be seen will remain true, that an accidental injury is but one factor in a *fixed sequence* … that this event can occur only because of either a positive or a negative unsafe action of some person or the existence of an unsafe physical or material condition; that this unsafe action or condition is created only by the fault of some person; and that the fault of a person can be acquired only through inheritance or through environment. [53; italics added]

There are a number of noticeable things about this statement, principally the lack of a social perspective. Person, for example, is singular. Even though Heinrich acknowledged that there might be situational features, this is kept at the level of the context surrounding an individual. Sociological or distributed factors are ignored. Moreover, this is very definitely a Newtonian view of reality. Events proceed deterministically via a fixed sequence. Human actions also are determined by things they have acquired through inheritance or environment. In other words, humans commit errors because of genetic factors or because of their upbringing.

One might argue that Heinrich was doing his best given the state of science in his time. However, his work is still in print, and as late as the edition of 1978, Peterson and Loos stated in the introduction (written after Heinrich's death): "It [i.e., Heinrich's work] was and still is the basis for *almost everything* that has been written in industrial safety programming from the date it was written until today" [54; italics added]. Perhaps even more surprising is the fact that behavioral safety is still practiced today.

Contemporary followers of Heinrich tend not to state (openly), as Heinrich did, that undesirable traits of character such as avariciousness are genetically programmed, and that these traits are important causal factors in accidents [55]. Nevertheless, they do tend to accept, along with Heinrich, that accidents follow the domino mechanism. That is, there is an initial causal factor that causes another factor, then another, then the accident, and finally the injury (like dominoes toppling). What is interesting (given that this is a Newtonian model that is supposed to apply in all situations) is the order of the events. Heinrich accepts that mechanical or physical hazards (what we would now call systems factors) play a part in creating accidents. But, these are always preceded by ancestry and social environment causing the fault of person. In other words, the individual is always the starting point for accident causality. To use contemporary jargon, according to Heinrich, the individual (or his genetic/learned defects) is always the root cause of the accident. Systems factors can only ever be an ancillary factor given that accidents "*invariably* occur in a fixed and logical order" [56; italics added]. Therefore, "man failure causes the most accidents" [57].

It is important to remember that Heinrich was led to this conclusion because he *began* with certain abstract laws and then went about looking for evidence to prove them (the view that Heinrich's findings were inferred from the data does not stand up to close scrutiny, as we will see).

Be that as it may, it is fair to say that Heinrich's major finding, that a majority of accidents are caused by human error, is one of the foundation stones of contemporary safety science, and that variations are still recycled in the literature (although the specific numbers cited tend to vary from author to author). Given this, it is important to note that Heinrich's findings have frequently not been replicated. For example, a study by Haggland found that between 54% and 58% of accidents were associated with systems factors, and only 26% to 35% were associated with human error [58]. Now, we are not (and given our philosophical biases could not be) claiming that Haggland's study is correct and Heinrich's is not. What we are merely saying is that it is interesting that other authors, looking at similar data, have produced completely different attributional paths. Haggland's study highlighted the *variability* in attributional styles and might be seen to indicate that Heinrich's choice of attributional explanation is in line with his attributional biases (which, to repeat, everyone has) rather than because it is a reflection of what is actually the case.

Haggland's study is interesting in that he made some salient points that are rarely discussed in the field. For example, he pointed out that "first report of injury is a poor source of information insofar as accident causes are concerned. It is almost always filled out by a management representative. … The reports almost never reveal an unsafe working condition as the cause of accident. … Generally the description of the accident consists of a short, terse sentence [59].

This is important. Haggland pointed out that the initial data are inadequate in many, perhaps most, workplaces, and any inference from these data should be made with care. More important is this implicit acknowledgment that attributions of causality are not disinterested (we have already seen this in the Woodcock and Smiley study). Management representatives seem to be mysteriously unwilling to blame themselves for accidents. Ipso facto, one might assume, if it was the staff/workers themselves who fill out the forms, perhaps different causes might be produced.

This can be made clearer by looking at an example Heinrich gave and the interpretation he offered.

1.4.1 HEINRICH'S EXAMPLE

Heinrich began by presenting a hypothetical example to back up his a priori assumption that the behavior of the person at the man–machine interface is the key causal factor in the majority of accidents. He presented the case of an employee in a woodworking plant who fractured his skull as a result of a fall from a ladder. Heinrich states that investigation showed that the ladder did not have nonslip safety feet. Moreover, the ladder was positioned at the wrong angle. Heinrich also says, in his usual inimitable fashion, that the man's "family record was such as to justify the belief that reckless and wilful tendencies had been inherited" [60].*

Now, it is clear that, if one is inclined to attribute such events to internal factors, one could argue that this accident was the worker's own fault. After all, the man was (born) a reckless sort, and he positioned the ladder wrong. On the other hand, if one's bias was toward external attributions, one could equally argue that the causal pathway began with (or can be traced back to) management. We know the ladder was inadequate, and one could argue that the faulty positioning of the ladder was a training issue. Or, in the final analysis, one could argue that, even if the man *was* incompetent, he should never have been hired in the first place (staffing is a management issue).

This seems self-evident, but the hard point to grasp is that *both* attributional accounts, which lead to different causal pathways, are biased and subjective and (presumably) have at least a modicum of validity. That is, there is no objective way of deciding which is the true causal pathway.

Another example might make this clearer. In the field of medicine we might ask the following question: what causes deaths by tuberculosis? Here, the answer seems to be laughably simple. Tuberculosis causes deaths by tuberculosis. And, when asked, "What causes tuberculosis?" most doctors could give a straightforward answer: the tuberculosis bacillus (*Mycobacterium tuberculosis*). What, therefore, should be done about this? The answer to the problem of tuberculosis, therefore, inferred from this particular causal chain is simple: eliminate the tuberculosis bacterium, remove this root cause. The invention and use of drugs like isoniazid and the development of the BCG vaccine are means to this end.

However, as Richard Lewontin showed, other causal chains are possible that look at other possible causes of deaths by tuberculosis. For example, in the 19th century (especially the 1830s, when the disease was particularly virulent), tuberculosis was particularly common and fatal among factory workers and the urban working class generally. Death rates were much lower among the aristocracy and the peasantry. Average death rates began to decline in the 1840s and slowly continued to decline for the next 130 years. Neither Koch's germ theory of tuberculosis in the 1880s nor the development of antitubercular drugs in the 1950s had any particular effect on death rates. By the time of the introduction of such drugs, the statistical rate of death by tuberculosis had already declined by 90% [61].

This clearly suggests a different causal chain and a different root cause, which leads to different intervention strategies. What this causal chain suggests is that the root cause of tuberculosis deaths is poverty. Certainly, the tuberculosis bacterium features in this causal chain, but as a consequence, or an activating agent, the straw that breaks the camel's back, in a person already weakened by years of poverty and misery. What this causal chain suggests is that the best way of reducing the statistical

---

* See Chapter 5 for a discussion of the accident-prone personality.

rate of tubercular deaths is to reduce poverty; sure enough, as the West has become richer, the death rate has come down.*

However, can we really say that this second causal chain is better (or, for that matter, worse) than the first? By what criteria? How would it benefit a laboratory worker busily studying the genome of the mycobacterium to think like this? On the other hand, how would a reductionist study of the DNA of the bacterium benefit a public health worker or a trade unionist attempting to improve the conditions of factory workers?

1.5   THE MYTH OF THE ROOT CAUSE

Needless to say, this questions the concept of the root cause and techniques of root cause analysis currently widely used in industry. The root cause analysis approach relies on the idea that if one follows the causal chain back far enough, one will eventually end up at one concept, fact, or event that functioned as the one primary causal element that caused all the others — the root cause.

But, there is a major problem with this viewpoint, criticisms of Heinrich notwithstanding. Think of the causal factors in the causal chain or series of dominoes waiting to topple each other. It is clear that the ones in the middle function not only as causes but also as consequences. For example, the middle causal factor is a causal factor and a consequence of the event that happened before it. It is also clear that in theory any event could function as either a cause or a consequence. For example, imagine a man knocked down by a truck. Now, we discover that the driver crashed because he fell asleep. So, now we decide that this is the root cause. However, we then discover that due to his shift patterns he had been unable to gain an adequate amount of sleep. The fact that he was tired (and subsequently fell asleep) is both a cause of the accident and a consequence of the fact that he had been overworking. But, why had he been overworking? Perhaps the firm had just been taken over, and new management insisted on new, tougher shift patterns. Or, perhaps the driver simply wanted the money because he wanted to go on a vacation, which invites the question of why he wanted to go on a vacation (perhaps this was because of the new management) and so on. In other words, no matter how deep the root cause, we can always treat it as a consequence and go one further back, on and on, presumably until the "big bang" and the creation of the universe.

What is suggested tentatively is that, in a real-world situation, perhaps the further back we travel in terms of investigating the accident, the more causal options we have (i.e., we have more and more data that we could, in theory, describe as causal), and that perhaps there are subjective criteria that might decide how this decision is made. Is it possible that the decision to see (or classify) something as causal or not is to a certain extent in the eye of the beholder?

---

* And there are further attributional choices available. For example, is it the poverty itself or the hopelessness and misery that go with the poverty that causes the deaths? Leonard Sagan argued that it is the lack of control and the powerlessness associated with poverty that lead to a weakened immune system and hence ill health. We return to the concept of lack of control in the final chapter [62].

1.6 MODELS OF ACCIDENT CAUSATION

Before we answer that question (or at least attempt to answer it), we should look briefly at a related problem: the search for the one true model of accident causation. Heinrich not only created the first proposed *methodology* for accident investigation, but also proposed the first *model* for accident causation. As we have seen, the search for models is closely associated with the search for laws, and the key aim of both is to create causal regularities that can act as the backbone for the new science. Heinrich again put himself right in the TR mainstream. Again, it must be asked, was this a wise move?

1.6.1 HEINRICH AND MODELS

Heinrich's model of accident prevention is the (still surprisingly popular) domino theory. In this view, which is openly deterministic, the causal factors fall like dominos. Therefore, there is one set root cause that triggers another and then another until the accident happens. This view is extraordinarily crude and simplistic, but it is arguably an improvement on the single-cause hypothesis, which simply looks for one single cause of any accident or incident.* Nevertheless, it has a number of nontrivial problems as a model.

First, we have the following problem: which is the first domino? And, who decides? That is, how far back do we follow the train of dominos? As we have seen, there is nothing obvious or commonsensical about this. The perception of a cause and how far back we look for the root cause are profoundly perspectival (that is, they vary according to the perception of the person looking for the cause) and are essentially up to individual investigators. It is little wonder that, in the field, investigators rarely agree with one another regarding the cause of any specific accident.

Perhaps an improvement on the domino theory is the multicausal theory, which views the causal pathways as more similar to a tree of multiple causation. This at least acknowledges the problems of emergence and synergy. It also acknowledges the polycausal nature of many events.

It should also be noted that there are other theories of accident causation, such as James Reason's latent pathogen concept [63] (the so-called Swiss cheese model) and many others (Benner noted, in terms of models of accident causation, "at least seven differing investigative processes, 44 reasons for investigating, six fundamentally different methodological approaches, three distinctive types of investigative outputs, and ... a total absence of criteria for establishing the beginning or end of the phenomenon being investigated" [64]). However, it must be questioned how different these views are, fundamentally, from Heinrich's original domino theory. All these views still tend to posit causes (or, in the case of Reason's theory, causelike entities) that have a deterministic relationship to one another. As we have seen, in terms of 21st century science these views are, strictly speaking, unscientific.

---

* This single-cause view still flourishes in the world of medicine, in which we are told that the almost infinitely complex dynamic system of the human body is stopped by a single "thing": heart attack, cancer, or whatever.

This is unfortunate because there is another accident model that avoids these problems. This is the multilinear events sequence perception model. Instead of causes, this views accidents as the result of interrelating processes. Moreover, these relations are not deterministic but are instead more like the arrangement of the individual notes in a musical score. (This difference is rather crucial. For example, think of the famous opening of Beethoven's fifth symphony: "da da da DUM." In a concert hall, the third da is always followed by the dum, yet one would never argue that this means that the da caused the dum.) As with a musical score, the accident is a systems effect, occurring in a system that has emergent properties. Moreover, the sciences used as a metaphorical paradigm are biological rather than mechanical [65].

However, even this model has a number of flaws. First, it still presupposes that one size fits all: there will be one accident model that might, theoretically, apply in all cases. Second, we still have the problem of reliability. Even if we can decide on the methodology, how do we make sure that two separate investigators would investigate the same accident or incident and produce the same results?

This occurs on two basic levels: intramodel and intermodel. Intramodel reliability is the problem, given the perspectival nature of causation, of how to know that two different investigators will select the same causes (or processes). And, if they do not, how do we decide between their two different "stories"?

Intermodel reliability is the same problem writ large. As again Benner pointed out, the concept of truth as this word relates to accident causes (or for that matter to accident causation models) is system specific. So, one system will provide one set of true causes, and another system will provide another. For example, a cognitive approach will tend to produce causes that include internal cognitive factors as causal elements. A systems approach, on the other hand, will concentrate on environmental or systems features, and so on. How can we decide which one is better or truer [65]? The only real answer is pragmatism, but this will only get us so far because we then have to decide the context in which it is pragmatic. What are these systems pragmatically useful *for*? Do they actually reduce accidents? There is little evidence that they do. Perhaps some are easier to use than others, but so what? If the answers they give are not true, then ease of use hardly comes into it. So, we come back to this problem again. If we do not have one model for accident investigation — if we do not have any laws for our science — if we cannot even agree on the simplest methodological questions, where is safety science going? What has it achieved?

In the next chapter, we start to look at some of these questions and, it is hoped, provide some answers. The answers, we believe, lie in the seemingly arcane study of taxonomies.

REFERENCES

1. Schön, D.A., *The Reflective Practitioner,* Basic Books, London, 1991, p. 21.
2. Shotter, J., *Images of Man in Psychological Research,* Methuen, London, 1975.
3. Burtt, E.A., *The Metaphysical Foundations of Modern Science,* Dover, London, 2003.
4. Ribe, N. and Steinle, F., Exploratory experimentation: Goethe, land, and color theory, *Physics Today,* 55 (7), 2002.

5. Schön, D.A., *The Reflective Practitioner,* Basic Books, London, 1991, p. 24.
6. Waller, J., *Fabulous Science,* Oxford University Press, Oxford, U.K., 2002.
7. Brown, A., *The Darwin Wars,* Simon & Schuster, Sydney, 1999, p. 47.
8. McComas, W.F., Ten myths of science: reexamining what we think we know about the nature of science, *School Science and Mathematics,* 96 (1), 10, 1996.
9. Marx, K. and Engels, F., *The Holy Family,* Languages Publishing House, Moscow, 1956, p. 125.
10. See, for example, Kuhn, T., *The Copernican Revolution,* Harvard University Press, Cambridge, MA, 1990, and Collins, H. and Pinch, T., *The Golem: What Everyone Should Know about Science,* Cambridge University Press, London, 1994.
11. Popper, K., *The Logic of Scientific Discovery,* Routledge, London, 2002.
12. Wittgenstein, L., *Philosophical Investigations,* Blackwell, London, 1999, p. 31.
13. McComas, W.F., Ten myths of science: Reexamining what we think we know about the nature of science, *School Science and Mathematics,* 96 (1), 10, 1996.
14. See, for example, Lakoff, G. and Núñez, R., *Where Mathematics Comes From,* Basic Books, New York, 2000.
15. Cartwright, N., *The Dappled World,* Cambridge University Press, Cambridge, U.K., 1999.
16. Ormerod, P., *The Death of Economics,* Faber and Faber, London, 1994.
17. Putnam, H., *The Collapse of the Fact/Value Distinction and Other Essays,* Harvard University Press, Cambridge, MA, 2002.
18. Putnam, H., *The Collapse of the Fact/Value Distinction and Other Essays,* Harvard University Press, Cambridge, MA, 2002, p. 19.
19. Cartwright, N., *The Dappled World,* Cambridge University Press, Cambridge, U.K., 1999.
20. Davies, J., Ross, A., Wallace, B., and Wright, L., *Safety Management: A Qualitative Systems Approach,* Taylor & Francis, London, 2003.
21. Winsberg, E., Laws and infinite regress: Puzzles in the foundations of statistical mechanics, presented at Karl Popper 2002 Centenary Conference, Vienna, July 3–7, 2002.
22. François, L., *The Birth of Mathematics in The Age of Plato,* Hutchinson, London, 1964.
23. Crowden, A., Ethically sensitive mental health care: is there a need for a unique ethics for psychiatry? *Australian and New Zealand Journal of Psychiatry,* 37, 143, 2003.
24. Vaughn, D., *The* Challenger *Launch Decision*, University of Chicago Press, Chicago, 1997.
25. Odling-Smee, F.J., Laland, K., and Feldman, K., *Niche Construction,* Princeton University Press, Princeton, NJ, 2003.
26. Edkins, J., *Whose Hunger?* University of Minnesota Press, Minneapolis, 2000, p. 20.
27. Schön, D.A., *The Reflective Practitioner,* Basic Books, London, 1991, p. 41.
28. Wilde, G., *Target Risk 2,* PDE Publications, Ontario, 2001.
29. Green, J., *Risk and Misfortune,* UCL Press, London, 1997.
30. Collingwood, R., *An Essay on Metaphysics,* Oxford University Press, Oxford, 1940, p. 304.
31. Kletz, T., *Learning from Accidents,* Gulf, London, 2001.
32. Hume, D., *A Treatise of Human Nature,* Everyman, London, 1974, p. 95.
33. Russell, B., *A History of Western Philosophy,* Unwin, London, 1988, p. 643.
34. Cartwright, N., *The Dappled World,* Cambridge University Press, Cambridge, U.K., 1999.

35. Hacking, I., *The Taming of Chance,* Cambridge University Press, Cambridge, U.K., 1987, p. 54.

36. Surry, J., *Industrial Accident Research,* Ministry of Labour, Toronto, 1979, p. 23.

37. Pattee, H., Causation, control, and the evolution of complexity, in Andersen, P.B., Emmeche, C., Finnemann, N., and Christiansen, P. (Eds.), *Downward Causation,* Århus University Press, Århus, Denmark, 2000, Ch. 2.

38. McCauley, C. and Jacques, S., The popularity of conspiracy theories of presidential assassination: a Bayesian analysis, *Journal of Personality and Social Psychology,* 37 (5), 637, 1979.

39. Fischer, D.H., *Historian's Fallacies,* Harper Torchbooks, New York, 1970, p. 177.

40. Miller, J.G., Culture and the development of everyday social explanation, *Journal of Personality and Social Psychology,* 46, 961, 1984.

41. Hamilton, V. and Sanders, J., *Everyday Justice,* Yale University Press, New Haven, CT, 1992.

42. Lillard, A.S., Ethnopsychologies, cultural variations in theory of mind, *Psychological Bulletin,* 123, 1, 1998.

43. Strauss, A., Northern Cheyenne ethnopsychology, *Ethos,* 5, 3, 1977.

44. Salminen, S. and Seth, G., Attributions related to work accidents in Finland and Ghana, *Psykologia,* 35, 416, 2000.

45. Gilbert, G. and Mulkay, M., *Opening Pandora's Box: A Sociological Analysis of Scientist's Discourse,* Cambridge University Press, Cambridge, U.K., 1984.

46. Guimond, S., Begin, G., and Palmer, D., Education and causal attributions: the development of "person-blame" and "system-blame" ideology, *Social Psychology Quarterly,* 20, 1, 1989.

47. Sullivan, D., *Psychological Activity in Homer,* Carleton University Press, Ottawa, 1988.

48. Woodcock, K. and Smiley, A.M., Organizational pressures and accident investigation, in *Proceedings of the Human Factors Association of Canada,* HFAC, Mississauga, Ontario, 1998.

49. James, R., *Return to Reason: Popper's Thought in Public Life,* Open Books, Chippenham, U.K., 1980, p. 13.

50. Heinrich, H.W., *Industrial Accident Prevention,* 4th ed., McGraw-Hill, New York, 1959.

51. Heinrich, H.W., *Industrial Accident Prevention,* 4th ed., McGraw-Hill, New York, 1959, p. xi.

52. Heinrich, H.W., *Industrial Accident Prevention,* 4th ed., McGraw-Hill, New York, 1959, p. xi.

53. Heinrich, H.W., *Industrial Accident Prevention,* 4th ed., McGraw-Hill, New York, 1959, p. xii.

54. Heinrich, H.W., *Industrial Accident Prevention,* 4th ed., McGraw-Hill, New York, 1959, p. viii.

55. Heinrich, H.W., *Industrial Accident Prevention,* 4th ed., McGraw-Hill, New York, 1959, p. 15.

56. Heinrich, H.W., *Industrial Accident Prevention,* 4th ed., McGraw-Hill, New York, 1959, p. 15.

57. Heinrich, H.W., *Industrial Accident Prevention,* 4th ed., McGraw-Hill, New York, 1959, p. 19.

58. Haggland, G., Causes of injury in industry, in Petersen, D. and Goodale, J. (Eds.), *Readings in Industrial Accident Prevention,* McGraw Hill, New York, 1981.

59. Haggland, G., Causes of injury in industry, in Petersen, D. and Goodale, J. (Eds.), *Readings in Industrial Accident Prevention,* McGraw Hill, New York, 1981, p. 19.

60. Heinrich, H.W., *Industrial Accident Prevention,* 4th ed., McGraw-Hill, New York, 1959, p. 18.
61. Lewontin, R., *The Doctrine of DNA,* Penguin, London, 2001.
62. Sagan, L., *The Health of Nations,* Basic Books, New York, 1989.
63. Reason, J., *Human Error,* Cambridge University Press, Cambridge, U.K., 1990.
64. Benner, L., Accident perceptions: Their implications for accident investigators, *Hazard Perception,* 16 (11), 4, 1980.
65. Benner, L., Methodological biases which undermine accident investigations, presented at The International Society of Air Safety Investigators 1981 International Symposium, Washington, D.C., September 1981.

2 Safety and Taxonomies

2.1 INTRODUCTION

In the last chapter, we questioned the idea that science is a single unitary thing with only one methodology. Instead, we showed that there are sciences (plural), each with its own philosophies, assumptions, and practices. Safety science, like all sciences, should adopt a pragmatic approach regarding methodology (or, rather, methodologies). So, the next question is, which methodologies should safety science adopt?

The best answer to that question, in line with our pragmatic bias, is to show how we went about some specific projects (specifically, arranging and organizing databases) and show what we did and why. Obviously, when we began, we started with the presupposition that gaining information is very important (not a terribly controversial point), and that one of the key tasks of safety science is to use this information in the most useful and economical fashion.* Again, this is not terribly controversial. However, keep in mind the various differences between the technically rational (TR) approach (which looks *through* the data) and the exploratory approach (which looks *at* the data), as it is hoped this will illuminate the difference between our approach and the standard approach to safety data.

At this point, many safety managers and safety scientists would wish to make a series of distinctions between sources of data that can be analyzed, between, for example, data from minor event reporting systems, confidential or whistle-blowing systems, accident investigations, root cause analyses, simulations, and so on. What we are going to argue here is that the source of the data does not really matter because the key point is that data gathered from whatever source and in whatever form (e.g., textual or numeric) have to be treated similarly: as information to be classified and analyzed. So, the source or form of the data does not matter. What matters is simply that you have data that are faithfully and rigorously gathered so that they are of sufficient quality to be analyzable.

However, one point must be cleared up before we start. There is one kind of data-gathering operation in safety science that attracts all the attention and is the best-known method outside the field. This is the classic accident investigation of a

---

* Specifically, we are looking at the analysis of texts. However, as we make clear in the chapter, the basic rules we are discussing apply whether raw data are textual, verbal, or pictorial (or, in fact, numerical, for which a taxonomy *has already been applied*). We thus argue that the theory we are discussing is a useful tool for almost all safety science. (In fact, we hope to say something relevant for all social sciences.) What this omits, of course, is the role of experimentation, which is certainly a useful tool. However, as we discuss in Chapters 4 and 7, perhaps its role has been overemphasized in this particular context. Experimentation has specific problems that mean that if experiments *should* be used, then a bit more caution than is usually taken is required.

major catastrophe, such as those involving Three Mile Island, *Challenger*, Bhopal, and so on. At the moment, we are going to push these specific grand-scale accident investigations to one side on the grounds that these are *not typical*. Most safety management does not consist of the investigation of large-scale disasters or incidents in which many people lost their lives. Instead, most safety managers spend their time analyzing smaller-scale accidents, incidents, and issues, whether one terms these routine events, minor events, near misses, or whatever. There is a good reason for this: general trends and patterns only become obvious when one deals with more than one data point, preferably as many as possible. And (this point seems obvious but will become increasingly important), it should be stressed that such data currently are normally held in a database of some sort. So, how should such databases be created? What should be done with the data held in them? The answer, we believe, lies in the study of *taxonomies*.

This chapter attempts to answer the question, "How should safety science be carried out given the philosophical conclusions of Chapter 1?" We begin by asking questions that are probably obvious to insiders but not so obvious to outsiders: what is the *point* of a database, and what are the advantages of having a taxonomy to arrange it?

2.2 THE PURPOSE OF A DATABASE AND A TAXONOMY

For many safety managers, the answer to the question, why have a database?, will be self-evident: because they have to have one. In some high-consequence industries (such as the nuclear industry), it is now mandatory to catalogue every safety event in the form of a short (or longer) report. However, even when it is not legally required, increasingly many industries are choosing to create similar databases of various sorts, some confidential (i.e., the person submitting the report remains anonymous to the organization), some anonymous (the person submitting the report remains anonymous to everyone and is untraceable), some nonconfidential (i.e., the report is submitted through official channels), and so forth [1]. The reason they do this is simple: they hope to gain information about what is really happening safetywise in their company.

The advantages of a database approach (as opposed to one-off accident investigations) are simple: a database enables the safety manager to look at many different safety events and look for patterns, such as similarities among accident types; accidents occurring in the same area (or part of the plant); accidents occurring at a certain time (e.g., 3:00 A.M.); and variations of these (e.g., it might be possible to see that a disproportionate number of reports concern events of a certain type in a certain part of the plant in the late afternoon; this would of course be a cue for further investigation that might look at shift patterns, what was going on in that area of the plant, or whatever). Moreover, with these data we can look for trends, such as the following: is the safety record getting better overall or worse? Or, is it getting better in some areas but not in others? For example, the sociologist David Philips and coworkers argued that, in the first few days of each month, fatalities due to medical errors rise by as much as 25% (he speculated that this is because workload increases at this point for various reasons). If this is true, as a safety manager this is clearly the sort of information you would want to know and act on, and it is clear that only by looking at a large number of cases could you draw these sorts of conclusions [2].

But, any way of interrogating a database presupposes that the safety data are classified in some way so that we can look for these patterns. In fact, the events *must* be classified; otherwise, the result is a completely unmanageable data set. We could go further and state that any database of any sort has to be organized in some way. Even arranging qualitative reports in alphabetical order is a kind of classification. So, the next question is, how do we go about organizing our database? This brings us to our work in this specific area.

2.2.1 STRATHCLYDE EVENT CODING AND ANALYSIS SYSTEM

The first major project we (as researchers at the Centre for Applied Social Psychology at the University of Strathclyde) carried out in the safety industry was with the British nuclear industry and was called Strathclyde Event Coding and Analysis System (SECAS). As the project began, representatives from the industry were concerned that their minor event database was not functioning in an optimal manner. It is important to realize that this database was *not* an "in-your-own-words" database. In other words, it was not a database in which purely qualitative data (transcribed from interviews) were stored (this is important because we are keen to stress here that the organizational methods we are talking about can apply to any database, not just free-text ones). Instead, the data were in the form of a short report of the event, which was, in essence, the report of a short accident investigation carried out by an experienced safety manager. Once this report (which might be as much as four or five pages long but was usually about a paragraph) was compiled, it was then stored electronically and coded for classification purposes by various coders. There was a long list of codes, which the coders would check to decide which ones applied. We were particularly interested in the coding part of the system. Here, coders might decide that an incident occurred due to a rule violation or a communications problem (or whatever). Two or three different people were responsible at any one time for this classification of events, and they would usually apply two or three codes to each event (i.e., each event would be classified in two or three ways). Another database was then compiled of the coded data so that, for example, data could be presented about the number of communications issues or work practice issues that had occurred in any given year.*

The problem that the industry had was that they were unclear whether the coding procedure was effective (and how this might be assessed), and that this was indicative of a more general problem of what to do with the data anyway.

2.2.1.1 The Reliability Trial

It was as a result of these concerns that the University of Strathclyde was brought in to carry out a reliability trial on the coding system. We go into more detail regarding how to carry out a reliability trial in Chapter 3 and the appendix. It is sufficient to say

---

* This is important because, as we discuss in this chapter, we have also worked on confidential reporting databases of precisely this type. The key point, however, is that the source of the data does not really matter.

TABLE 2.1
Interrater Reliability Data for Original Nuclear Root Cause Event Coding System (n = 28 events; 4 coders)

Paired Comparison	Coders 1 and 2	Coders 1 and 3	Coders 1 and 4	Coders 2 and 3	Coders 2 and 4	Coders 3 and 4	Average
Index of Concordance	42%	45%	41%	42%	41%	39%	42%

at the moment that a *reliability trial* (in our use of the phrase) measures the degree of consensus that exists between coders in terms of the coding or classifying of individual events. Accordingly, we had experienced coders code a series of randomly selected reports from the database (in separate rooms, with no conferring) to see the extent to which they agreed with one another on how they would classify or code them. Why did we do this?

The short answer is that if we do not have consensus in interpretation in individual events, then clearly the amount of noise in the database will quickly increase to the extent that the database will become valueless (we provide specific examples of this in Chapter 3). Or, to put it more clearly, if we are going to classify the event in a database, we have to show that *same* event would be classified in the *same* way by *different* people. If the same event would be coded or classified differently by different people (or differently by the same person after some time had elapsed), then it is clear that locating specifics of events will become impossible; moreover, counts of events will quickly become meaningless. If I classify an event as type A, and you classify it as X, then when I count how many A events there are, this will depend on whose classification I use. We really want frequencies of events to say something about the system rather than about who happened to be doing the classification at a particular point in time.

Twenty-eight previously coded events were randomly selected from the database and were presented to three experienced coders or classifiers. They were told to code the events in the usual way. They were *not* allowed to discuss the events either before or during the trial.* The events had already been coded in the database, and these original codings or classifications were used as if they were the codings of a fourth coder. Accordingly, there were six possible paired comparisons for measuring reliability (consensus). The results of this trial are given in Table 2.1 (note that reliability was provided in this case by the use of the index of concordance [3]).

It can be seen that average reliability in coding was 42%; that is, the coders were only applying the same codes to the same events about 42% of the time. It is generally agreed that an acceptable degree of consensus in trials of this sort is 70% [4]. It is clear (or should be) that those in management were right to feel that the system was not functioning in an optimal manner.

---

* Discussion was allowed after coding had taken place.

2.2.1.2 The Problem with the Codes

An obvious solution to this problem would be some sort of additional training. It could be argued that training in using the system was inadequate, but these were experienced coders/classifiers who already had extensive training and experience in using the system. So, what other problems might there be?

Asking this question led us to look at the coding system (i.e., the lists of codes or classifications) itself. In this system, there were 196 codes/classifications distributed among 17 supercategories (with titles such as management methods and design, configuration, and analysis). Moreover, the codes had evolved in an ad hoc manner (rather than having been derived from a coherent scheme that modeled action in the industry). There were differences in the way the codes or classification were interpreted between plants.

In other words, it seems that there was a problem with the codes themselves or, to be more specific, with the *taxonomy* (a word meaning hierarchical classification or ordering, from the Greek *taxos* [arrangement] and *nomie* [method]) that ordered the codes. The rest of the project (which eventually was called SECAS for Strathclyde Event Coding and Analysis System) was dedicated to creating a new coding or taxonomic classification system that was easy to use and led to higher reliability or consensus in classification.

2.3 THE PRIVILEGED CLASSIFIER

At this point, we must clear up a common misunderstanding. It would seem like common sense (and, we would argue, it in fact *is* common sense) that what we must have before we have an adequate database of this type is consensus in classification. It should be common sense that if one was to reject this and simply have one person appoint himself or herself as an expert classifier, then the whole project would fall to pieces. In other words, you cannot have someone saying: "Well, I don't care what the consensus is. I am going to classify this thing as X [or perhaps as Y, or perhaps as X on one day and Y on another], and I am just *right*, and you all have to follow what I do." To repeat, this would seem to be contrary to common sense; yet, very surprisingly, some people do claim this. This is the idea of the privileged classifier.*

Before we continue, we must show what is wrong with the idea of the privileged classifier and why consensus (reliability) *between* coders is so important in the sense we have defined it, that is, that coders, when separated, can spontaneously agree on the classification of individual reports.

In the interests of clarity, it is probably going to make things clearer if we temporarily move away from the safety field and instead look at the general purpose of the project, to create a database. With any database, information has to be put into it and information taken out. Moreover, it is obvious that with any database of more than (say) five or six items, we are going to have to have some organizational aspect to it; in other words, we are going to have to arrange or order the database

---

* Granted, such persons would usually not refer to themselves in these terms. Instead, they would term themselves "expert classifiers" using "expert judgment."

so that (a) when we put an item in the database we know where it is when we wish to retrieve it and (b) we can keep track of how many items of particular types are in the database (for trending, cross-tabulating particular events with aspects such as time and place, etc.).

Unfortunately, people tend to become confused by the use of the word *database* and tend to think in terms of (for example) the World Wide Web. That is not really the type of database we are talking about here. The World Wide Web is a dynamic database; in other words, the mere act of looking for (and finding) something changes the odds of finding it again. The more popular a website is and the more sites it links to, the more likely it is to be selected by a search engine when we are looking for it the next time.

We are not talking about that kind of database. Instead, we are talking about a database in which we strongly want the data to stay in the same place. Moreover, this kind of database does not have to be electronic; perhaps it will simplify matters if we use the example of a library.

2.3.1 CLASSIFYING BOOKS IN A LIBRARY

For the sake of argument, let us take the example of a large public or university library with three librarians (although the same principles apply regardless of the size of the library). Now, we have a large number of books, and it is clear why they have to be classified; otherwise, nobody will be able to find anything! One more very obvious conclusion follows from this. There is no one correct way to classify these books. We could classify them by size, color, or anything. What matters is that we have a pragmatically acceptable way to categorize the books.*

However, let us assume that this library is sufficiently large that we wish to categorize the books by subject. Again (for the sake of this thought experiment), let us assume that the librarians have no knowledge or experience of working in a library or of any other mechanism of storing knowledge. All that we presuppose in this thought experiment is that the taxonomy (classification) of books has already been decided, and that this taxonomy is publicly available in a library catalogue (the taxonomic rule book), easily accessible to all the librarians.

So, on their first day of work, our librarians are faced with row upon row of empty bookshelves until the truck arrives with the first delivery, and a huge pile of books, magazines, tapes, and the like is dumped on their desk. So, with a sigh, our first librarian picks up the first book from this huge and muddled pile. Let us just say the book is entitled simply *Introductory Psychology*. Our librarian has never heard of the word *psychology* (many safety managers may be envious of this), so it is clear the librarian has to pick up the classification catalogue, which the library owners have helpfully provided. The librarian looks under P for Psychology and sees that books that contain the word *psychology* are to be put in the Psychology

---

* Similarly, there is no paradox or contradiction in having one library using one classification system and another using a completely different classification system. In fact, this is usually the case in the real world.

section of the library. So, the librarian looks up the floor plan, finds Psychology, and puts the book on the shelf.

So, we have one book on the shelves. Now, for the sake of argument, let us just assume that the second librarian has also stumbled on a book named *Introductory Psychology*, and this librarian also makes the journey to the catalogue and reads the same tautologous definition. However, whereas the first book was written by a Professor Abrams, the second book is written by a Dr. Zenovsky. So, what next? Again, the librarian consults the catalogue and sees (given that there is already a book of this name on the shelf) that the books are to be stacked in alphabetical order. So, the librarian places it *to the right* of the first book.

So far, so good. But, it should be recognized that the only difference between classifying something under Psychology and classifying it alphabetically is that the librarian did not have to consult the catalogue to remember the letters of the alphabet. The librarian knew them already (we assume). The breakdown of the signs and symbols commonly used in the West into 26 categories is, simply, another taxonomic categorization. Again, it is not privileged. In China or using the Arabic alphabet, there are different ways of categorizing sounds and symbols. The only difference is that (in our language) there is a high degree of consensus (in the West) in using this taxonomic categorization. Almost everyone, when asked to put books (or anything) into alphabetical order would get it right. We all know what the letter A, the letter B, and so on look like.

However, to return to our example, the next book in the new books pile to be picked up is entitled *Physics for Beginners*. The third librarian (again, never having heard of physics; it should be stated at this point that of course this is merely an example, and irate letters from librarians wishing to share their knowledge of physics and psychology will not be opened, let alone answered) looks in the catalogue or rule book and discovers the rule that books with the word physics in the title are to be placed under the category Physics. Things progress in this way for the rest of the day.

Let us recap before the bewildered reader decides that this is really a book on library management masquerading as a safety text. We now have two categories, Psychology and Physics. But, we also have a *subcategory* of *alphabetical ordering*. The fact that alphabetical categorizations do not automatically seem taxonomical (because we all know the alphabet) is merely because we have a very high degree of social consensus in applying this particular taxonomic categorization. But, for the purposes of this library, alphabetical categories are subcategories of other categories (which happen to be of subjects, i.e., psychology, physics).

Now, to return to our earlier point: why bother with this rigmarole in the first place? The answer should be obvious. First, when we have data, it has to be organized because otherwise people could not use the library (database): they would not know where anything was. So, there have to be categories or classifications, and these have to be public because we cannot assume that everyone knows what categories mean (or how they are defined by everybody) as in this case, in which the poor librarians do not even know what the words *psychology* and *physics* mean.

There is another reason the definitions must be public: *readers who come to the library have to know where the books are as well.* If, for some mysterious reason, a reader comes along who does not know that books with the word *psychology* in the title will be stored in the Psychology section (if, for example, the person thinks that this is one of these mysterious libraries where the books are categorized by color) and does not have access to the catalogue (rule book), then the person will leave bookless. If that is happening regularly, what is the point of a library? (Access to the catalogue usually does the trick, but remember the catalogue is a reliable way of finding books if and only if people use and interpret it in the same way.).

There is another reason why we want these definitions. Say that there has been a move toward the 21st century in this particular library, and a new electronic database has just been installed. Now, what would we want to know from this database? We want to know not only where books are on the shelves, but also how many books we have and how many books of each type (e.g., 20 physics books, 30 psychology books, etc.).

Unfortunately, even the best-run libraries also sometimes lose books. Say we had 20 copies of *Introductory Psychology* by Professor Abrams three years ago, 17 last year, and 15 this year. We obviously notice a trend, and we know that we have to buy at least 2 copies (and preferably 5) to make up the numbers. Again, none of this would be possible unless the classification was done correctly.

Finally, at the risk not so much of belaboring the subject as of beating it to death, what do we mean by correctly? We mean that the classifications must be known such that Librarian A, Librarian B, and Librarian C would all classify the same book in the same way, and that Reader A could, without speaking to any of them (but merely by consultation with the catalogue) find the book he or she was looking for in the right place.

Therefore (to translate this example back into the language of taxonomies), Librarians A, B, and C must have a high degree of social consensus in terms of their knowledge and use of taxonomic classifications. This is just a fancy way of saying that they all have to put the same book on the same part of the same shelf, year after year after year.

2.3.2 THE MYTH OF THE PRIVILEGED CLASSIFIER

It should now be obvious why there cannot be such a thing as a privileged librarian. You cannot have Librarian B rebeling against a lifetime of servitude and suddenly deciding in a fit of madness that psychology books must now be classified under Physics or that Z does not really work at the end of the alphabet and would look much better between E and F. This would lead to chaos.

Except, there is a catch. Say that this librarian did in fact decide this, but then persuaded the other librarians to follow this decision. Then, of course, the decision would be okay. As long as we have *consensus,* we have *order.* All that is important is that the readers also know this, so alterations must be made in the catalogue as well. As long as the catalogue is public and everyone knows they have to consult it, there cannot possibly be any problem with that.

So, given that this is the case and that it would seem irrefutable, why have some people argued that there *are* in fact privileged classifiers, that is, classifiers who cannot get people to agree with them (discussed in more detail in Chapter 3), but classifiers who are right in their classifications *even if most people disagree with them*? To answer this question, we must unfortunately dive into the murky waters of philosophy. However, it is worth doing, and it is hoped that at the end of this journey the waters will be a lot clearer than they were before.

2.4 THE CORRESPONDENCE THEORY OF CLASSIFICATION

To realize why some people have argued against what we are arguing here, we have to remember that taxonomic classifications are ubiquitous. In fact, one may describe human beings as Classifiers. That is not to say that animals do not also make taxonomic classifications, for example, the classification of hot and cold, good to eat and not good to eat. But, we are exceptional in that, because we have language, we can make far more complex and nuanced classifications than animals. When we enter our offices, for example, we can immediately make the taxonomic classifications of chair, desk, computer, book, and so on and so can everyone else with whom we work. It is obvious that if we could not all make these social categorizations, then social life would break down. We could go further and say that social life to a very great extent consists of taxonomic classifications, and it is the *social* aspect of these classifications we would stress.

There is one very important point that must be made here: we have at our fingertips a very large range of taxonomies, which we use in certain situations (and not others). In other words, the use of taxonomies is *context specific*.

Take the object of a four-legged animal (assuming for a moment we understand the taxonomic categories of four, leg, and animal) that goes woof woof. How do we categorize it? As the social psychologist Roger Brown was one of the first to point out, we actually have a large number of choices. We could call it "not only a *dog*, but … also a *boxer*, a *quadruped*, an *animate being*; it is the *landlord's dog*, called *Prince*" [5; italics added]. All of these are taxonomic categorizations, and they are all accurate, as far as it goes. In turn, they are all subdivisions of other taxonomic distinctions, which stand in various relations to each other (e.g., we could argue that boxer is a subcategory of dog and that quadruped is a subdivision of animate being). Even the landlord's dog and Prince are taxonomic categorizations. We always have at our disposal many taxonomies, which we can use as and when we please. When do we use one rather than another? As Brown put it, "Many things are *reliably* given the same name by the whole community … the most common name for each of these categorises them as they need to be categorised for the communities' non-linguistic purposes. The most common name is at the level of usual *utility*" [6; italics added]. In other words, we, reliably, as a community use certain taxonomic categorizations because they are useful to us as a community. We are rarely misled when we use the taxonomic categorization spoon because we all know what a spoon is, and we know what it is because we need it to eat our soup (or whatever).

It is important to note that other cultures do not have our social organization and therefore would not know which taxonomy to use in this context. In fact, the sheer range of taxonomies available is startling, as anthropologists have revealed to us. We think that, for example, the standard breakdown of our color taxonomy into red, orange, yellow, green, blue, indigo, and violet must be fairly universal, but it is not. For example, the Dani (a tribe from New Guinea) categorize *all* colors into two categories, light and dark. They simply have no concept (categorization) of the taxonomic categories with which we are familiar (red, yellow, etc.). Another tribe has a taxonomic category warm color, which includes red, yellow, and brown, and so on [7].

This view would seem to be the common sense one, and it is. But, surprisingly, some philosophers have propounded an alternative view, which they term the correspondence theory (although they do not use the language of taxonomies).* It is (we would argue) because of vaguely remembered ideas from philosophy such as this that the myth of the perfect taxonomizer has evolved.

The correspondence theory has, as it were, two aspects, a philosophical or epistemological component and a psychological one. We deal with the psychological aspect in Chapter 6. We deal with the epistemological view next.

2.4.1 PROBLEMS WITH THE CORRESPONDENCE THEORY

The correspondence theory states that there is one correct taxonomic classification for each individual item, and that therefore one person may possess it.** If I wished to categorize something as X, then I could in theory *just be right* no matter how many people disagreed with me (this belief is termed *essentialism*, the idea that items have an essence [which we would term a *fixed taxonomic classification*] that defines the taxonomy to which they naturally belong). But, correspondence theory will not work for three very important reasons.

The first reason is the variety of taxonomies we have available. Correspondence theory presupposes that we all use only one taxonomy. As we have seen, we do not. To use the classic example, as stated by Tarski: "Snow is White, if and only if Snow is White," which we would translate as "snow is to be categorized correctly as coming (at all times) under the categorization White if in fact it belongs to the taxonomic category White." In other words, snow is one of the many things that fall under the taxonomic category Things That Are White, if not at all times, at least in this [8].

What could be more commonsensical than this? But, remember the lesson of hermeneutics: all words, sentences, and behaviors are inherently polysemous (have many meanings). So, in this case, snow does not *just* mean frozen water, but also (so we have been told) might mean cocaine. Likewise, white does not just mean the color; it also means Caucasian. So, a perfectly reasonable translation of the statement above is "Snow is white if and only if cocaine is Caucasian!" which hardly gives us faith in the correspondence theory [9].

What has gone wrong here? The problem is that there is no answer to the question "what does this word (or sentence or behavior) mean" unless we specify *which*

---

* Instead, they would talk about the correspondence theory of truth, often taken to support metaphysical realism, a philosophical theory that is *not* the same as the common or garden realism that ordinary people use to live in the world, despite many claims to the contrary.

** It is normally said to be stored internally as a mental model, hence the link with psychology.

taxonomic arrangement we are using at the time. For example, white is a color under the taxonomic categorization Colors but a race under the taxonomic arrangement Race. Likewise, snow has one meaning when it is classified under the taxonomic supercategory Forms of Water and quite another when we classify it under the taxonomic supercategory slang Words for Drugs.

For that matter, even the rock bottom common sense idea of standing in front of snow and saying, "Look, I mean THIS white! And THIS snow!" will not work. The idea that it might was refuted by the philosopher Wilfred Sellars many years ago; Sellars pointed out that to state this is to presuppose that we all share the same socially acquired meanings of the words *white, snow,* and so on. This view he called "the myth of the given" [10]. In any case, this concept presupposes that two people looking at the same color, even without special training, will *see* the same color. Ethnographic and biological studies of how animals (and humans) see color seem to argue against this [11]. So, even stating "I mean THIS white" will not necessarily work ("but that is not white, that is what I call gray" might be the response).

Moreover, to some scientists, snow is really transparent, or even rainbow colored, and it is perfectly possible to imagine John standing in front of the window and saying, "Snow is white," to be corrected by Jill, who says, "Well, actually, it's not, it's transparent"; this debate might go on forever. So, who decides? The key point is that no *individual* can decide without using methods of persuasion (i.e., rhetoric) to attempt to build consensus. The idea of a privileged classifier is incoherent. John cannot gain authority because we cannot reach out of our phenomenological worlds and touch the real world, then compare it to our internal cognitive representation of snow, and then claim, "Look, these match!"* If we cannot do that (and we cannot), then just because John says, "Snow is white," does not make it so, no matter who he is (and it does not make his statement true; true and false are taxonomic categorizations like any other).

This is clearer if we look at the phrase "Snakes are reptiles." Of course, snake is also a term for a slimy, unpleasant *human being* (as in "He's a snake, a real creep"). You could argue that this view is metaphorical. But, as Lakoff and Johnson argued, the taxonomic distinction between metaphorical and nonmetaphorical language is almost impossible to make in real-world situations [12]. In any case, the point remains: we would have to know the context (i.e., which kind of taxonomy we were using) before we could decide whether this was the correct use of the word *snake.* Taxonomic distinctions are regulated by the social rules in which the taxonomies in which they function are used. Change the context (and therefore the rules), and you change whether the use is right or wrong [13].

To move to the second reason, is slush snow? What about ice? What about black ice? Or, black snow? Like all categories, Snow (and, for that matter, White) is fuzzy (i.e., *analog not digital*) [14]. This is in fact typical. We create distinctions everywhere, but the points at which we draw these lines are, to a large extent, arbitrary.

---

* This is not because our worlds are inherently subjective or anything like that, but because the taxonomic distinctions on which this picture relies ("inner," "outer," "objective," "subjective") are not useful in this context.

Again, what matters is the degree of social consensus about where the lines are drawn and how well these social rules are known. This is another reason why standing in front of the snow and saying, "Well, this snow and this white" will not work. Things not only are fuzzy, but also different people draw the digital distinctions that override the analog phenomena differently. For example, at what point does red shade into maroon? Would you and a friend both (under scientific conditions) identify precisely the same place? Are you sure? What if the light is poor, and you are in the light part of the room and your friend in the dark? What if your friend is shortsighted or partially color blind, or if your friend speaks Yélî Dnye, which not only has no word for orange or maroon (among other colors), but also has no specific word for color at all [15]?

We referred to these issues as fuzzy. This was a reference to a new branch of mathematical logic termed *fuzzy logic*. Unfortunately, fuzzy logic is often misinterpreted as a theory of epistemology (thereby leading to confusion between fuzzy logic and probability theory*), but in fact fuzzy logic is a theory of *ontology*.** In other words, just as purple shades into maroon, which fades into red, which shades into pink, but we classify these things into *discrete classifications*, fuzzy logicians argue that *all* our classifications are like this, blurring into each other: analog, or fuzzy, not digital or consisting (naturally) of discrete categories.

Fuzzy logicians would ask the question, what is the *real* distinction between a dog and a wolf? They would answer it by saying there is no *real* distinction. We (i.e., human beings) draw sharp distinctions, but that is because we have to (in this case, so our legs are not bitten off); what we cannot do is assume that these classifications are in some sense metaphysically true (remember the fact–value distinction, for which the same point was made; facts and values are an analog phenomenon on which we humans *impose* a digital or binary distinction into facts and values).

Particularly relevant here is the fact I might draw the line (in the above example, between a dog and a wolf) in a slightly different place from you. We discussed this briefly in terms of colors (red and maroon). To change the example for a second, if I say that it is raining, and you say it is not, it may simply be that where you draw the line in terms of the classification of rain might be different from where I draw it (e.g., one drop every 30 seconds might be defined by you as rain but not by me) [16]. So, genuine contradictions about classification are conceptually possible. This is not an insuperable problem because what we *can* do is build up and then measure our degree of consensus in drawing these distinctions. For example, if we get together and decide that two drops of water from the sky every 10 seconds is now to be classified as rain and fewer drops every 10 seconds is not, we have now solved the problem because we have a consistent definition of the category of rain, consistent because it is socially agreed on and used consistently by both of us. The water drops from the sky at the rate of one drop every 10 seconds: is it raining? We can now look up our shared definition and say, definitively, no. The fact that this line or definition is to a certain extent arbitrary (we could define rain in any way we want)

---

* In fact, this distinction is easy to make. Probability deals with issues pertaining to the future, and fuzzy logic pertains to issues pertaining to the present.

** Uncoincidentally, ontology is sometimes now used to refer to the organization of a database.

is irrelevant from this pragmatic viewpoint. Our new definition of rain is *reliable* (discussed in more detail further in Chapter 3).

For the third reason, finally correspondence theory states that the categorization (often conceptualized as like a picture, perhaps, stored internally, in the mind) corresponds directly to the object or thing outside. To which the only answer can be, well, says who? The statement "Snow is white" did not just come out of the air or from the heart of a burning bush. Some real person with lungs, teeth, and lips must have said it (or must have written it). In other words, there is no such thing as an abstract statement in this sense. All statements/sentences (and therefore all taxonomic categorizations) are made by real people of flesh and blood in specific situations (i.e., they are context specific). So, the statement with which we began this discussion should be glossed as follows: "Snow belongs to the taxonomic classification White *if John (or whoever) says it does,*" which seems to be conceding quite a lot to John. How would HE have privileged access to knowing the true taxonomic classification of anything? By definition, even if he does claim to have such knowledge, then this can only remain a tautology ("Well, I just *know*") unless he can persuade others of this claim. His claim would only work if he had some form of metaphysical God's eye (i.e., context-free) view of the matter, but only God can have that. We do not, because the decisions and classification decisions of human beings are all context bound. And, we must remember (having abandoned induction and the fact–value distinction) that all taxonomizers are *intrinsically* biased because all people are intrinsically biased. There is no such thing as *metaphysical* objectivity of the sort posited above (i.e., context-free objectivity). Therefore, if the observer manages to persuade everyone that the rules of correspondence are correct and that the leaf is green or that snow is white, then that is fine, but all that it shows is that if one analyzes correspondence theory closely, then in actuality we end up with the social consensus theory we are proposing.*

2.5 TAXONOMIES AND SAFETY

We had to deal with the correspondence theory to show that the idea of an individual taxonomizer is meaningless. Instead, taxonomies and categorizations are inherently and irreducibly social, and they only function in a social context.

Safety taxonomies are no different: what we must build up is social consensus in the use of taxonomies *that are specific to specific situations.* There cannot ever be "just the one" taxonomy.

Taxonomizing is a task, a practice that people carry out in the real world. If that is the case, then it is a skill that can be learned, the same as any other kind of skill: it is not in any sense true that we are born knowing it.**

---

* The correspondence theory (of truth) reached its "tightest" expression in the logical atomism of Wittgenstein's tractatus philosophicus. Notoriously, of course, Wittgenstein later abandoned this theory, and we are alluding here to his opposition to theorizing, as described in his *Philosophical Investigations*: the idea that no real philosophical theorizing is possible, only a description of the linguistic facts.

** In other words, we would argue that there are *no* hardwired taxonomies in the human brain. The idea that there is such cognitive hardwiring is one of the foundations of the psychological theory of cognitivism. For a further discussion of this, see Chapter 5.

So, how can it be learned? We are here proposing that there are criteria for the success or failure of taxonomy, and that the principal criterion is the reliability with which it can be applied or the degree of social consensus in applying it to individual cases. Only then can it be useful in discrimination and prediction (see Chapter 4). How can reliability be achieved? Let us return to the example of SECAS.

2.5.1 TAXONOMIES IN THE NUCLEAR INDUSTRY

To return to the nuclear project, it was clear that with the low level of reliability obtained in the use of the taxonomy, it would be necessary to start from scratch. It would, of course, have been possible to adapt some other taxonomy that had been used in some other industry. But, taxonomies are situation specific. Conceptually, as we have shown, there can be no one-size-fits-all taxonomy. Taxonomies should always be built with the needs of the specific situation in mind.

The key problem with the existing database was that the codes had evolved over time and had not been derived from a coherent logical structure. Therefore, to begin, after extensive interviews with management and staff (as well as numerous visits to nuclear power plants), we created a model of action in this industry (see Figure 2.1). This was based on a very simple three-way matrix in which a common sense distinction of actions having a beginning, middle, and end (or, in the terminology, actions were planned, implemented, and then monitored) was cross referenced with managerial levels of the plant. Actions at the managerial level were termed *distal*, actions at the supervisory (middle management) level were termed *intermediate*, and actions at the front line were termed *proximal*. There were also action codes that dealt with aspects of the event itself (specifically, whether procedures were violated) and monitoring codes that dealt with how the event was detected.

It should be stressed that this model was created after prolonged investigation of various NPP (nuclear power plants) (which involved physically going there and visiting them) and interviews with staff, management, and so on. In other words, it was specific to this particular industry. This is the key point to be understood: the codes were inferred from a model that was industry (or system) specific, and we would argue, this is the only way to create such a coding system (because how can you code what went wrong unless you know what *might* have gone wrong?).

To repeat, any system of codes, or coding taxonomy, must be system (in this case industry) specific.

2.5.2 SECAS CODES

The actual codes then had to be created. It should be noted that the codes not only were inferred from the basic logical structure of Figure 2.1, but also were created over a period of months in discussion with the people who actually had to use them. Specific issues (such as the use of subcontractors) were salient in this industry; again, this brings home the idea that taxonomies are situation specific.

After intensive discussion, it was felt that the best way to create the codes and present them would be to present them as a kind of logical flowchart. That is, the classifier would read the report and attempt to classify aspects at a fairly abstract

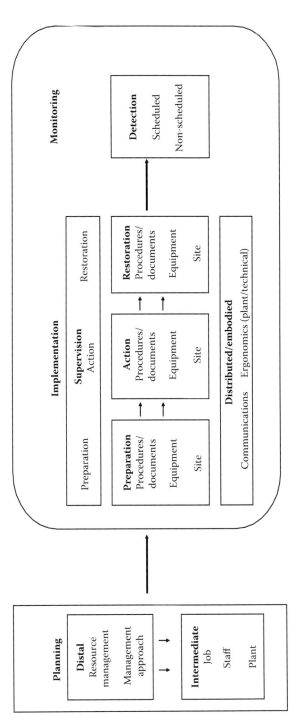

FIGURE 2.1 Model of action for the nuclear power industry.

level, progressing through coding decisions (or gates) to more and more concrete levels. So, if we take an example in Figure 2.2, we can look at the information transmission (communication) codes at the proximal level. The first stage, then, was to identify from the report whether the issue had anything to do with information transfer. If it had, then the classifier was faced with a choice: did it deal with information transfer between people (personnel communications), or from machine to person (information from equipment/plant)? If the former, then was the problem the *form* of the communication, the *content* of the communication, or the fact that there was *no* communication? Similar choices exist if B320 was picked.

Importantly, the classifier could stop at any time if there were insufficient data to proceed further, so that *some* information was coded for each event, if not all the detail. If the report did not make clear which code should apply, then the coder could code the event as a B310 or even a B300. This ability to match the detail in reports with the abstractness of codes is important for reliability of coding, and classifiers were taught that they should only proceed from left to right as far as they felt they were justified in doing so from the text available. This is only possible with a hierarchically arranged taxonomy.

2.5.3 Mutually Exclusive Codes

As can be seen from the example, the codes were nested and mutually exclusive. That is, as we move back from right to left, we can see that all the codes add up to the codes on their immediate left. So, B310 (Personal Communications) is made up of the three codes B311, B312, and B313: Content of Communication, Form of Communication, or No Communication, respectively. Logically, there are (or should be) no other possible codes. As we move back again, we can see that communication issues (in this particular plant) consist of those between people or between technical devices and people; logically, there *cannot possibly be any other kind of communication issue.** This use of the word *nesting* is taken from cladistics (in biology), in which, for example, the supercategory of Living Things is broken down into Animals and Plants; Animals is broken down into Amphibians, Reptiles, and Mammals (etc.); and so on down to individual species.

Moreover, not only is this hierarchy nested, the individual codes are also mutually exclusive. For example, a single problem must be definable dealing with the content of the communication or the form of the communication, not both. It is possible for a communication to be wrong in both ways, of course, and then coders have to agree on a pragmatic approach, for example, to code the most salient feature or to code both problems by treating this case as two events to be coded. All the codes at all levels have tight definitions, which are given in the coding manual (remember our example of the library catalogue). The manual must be freely available to all coders/classifiers, with examples that make clear the mutually exclusive nature of the scheme.

---

* Communication problems with information transfer *between* technical devices would be a technical issue for which there is a separate taxonomy.

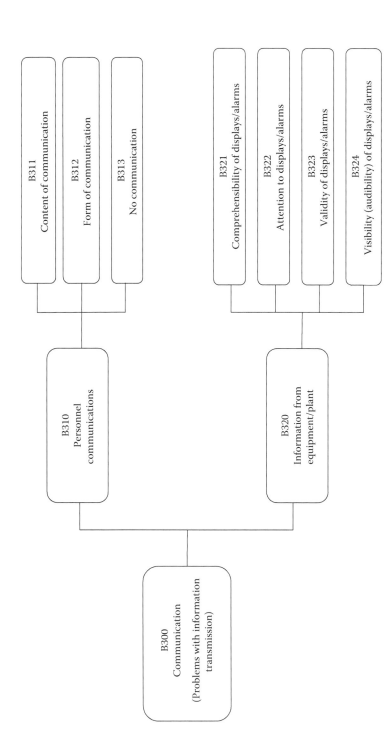

FIGURE 2.2 Hierarchical taxonomy of communication codes.

So far, so good. But, why bother? Why bother having nested, mutually exclusive categories? The short answer is that having taxonomies that are mutually exclusive and exhaustive (MEE) is generally agreed to be a good thing in terms of creating workable databases [17]. However to explain why, we have to return to our library example.

2.5.3.1 Mutually Exclusive and Exhaustive Categories

Let us assume that the library has a sufficient number of books such that the taxonomic category Psychology also has to be subdivided into further subsections before subdivision into alphabetical taxonomic categories. We subdivide psychology into cognitive psychology, social psychology, behaviorist psychology, and so on. Now, this sounds fine until we discover a book entitled, *Behaviorist Social Psychology.* Where is it to be classified? We could conceivably classify it as either a book on behaviorist psychology or a book on social psychology. Which is it to be?

Obviously, there is no natural grouping of this sort. In other words, for our pragmatic purposes there is no answer to this question. All that matters is that a decision is made, that it is known to all the classifiers, and that all the classifiers will reliably use whichever public rule is made available. For example, we may decide that in confound situations like this, whichever word is first is to be the deciding factor. In this case, we might decide that the word *behaviorist* comes first, so we decide that it should be classed under behaviorist psychology.

Again, we should realize that all taxonomic categories are in theory fuzzy like this; that is, there will be some books (or items) that could, in theory be classified under more than one taxonomic classification. The only way to deal with this situation is to have socially agreed on rules of taxonomic classification that are publicly available.

Now, the next point to grasp is that there is more than one way of classifying books in a library. Certainly, we may do it via subject, in the relatively crude way above. On the other hand, we may also decide to classify them in some other way, for example, via the Dewey (decimal) system or some other system. Again, it does not really matter as long as the classification system works. Also, if the library is of a reasonable size, it will be pragmatically easier if the taxonomy is hierarchical.

So, given that these are the aims of a good taxonomy, what are the rules for the creation of a good taxonomy? It is generally agreed that taxonomies should be MEE. In other words, each taxonomic category should exclude each other category, and all items that are required for the database should be included in the taxonomy (it is no good having a library with books in it that cannot be classified).

The importance of mutually exclusive categories is driven by the fact that, paradoxically, categories are never entirely exclusive of one another in reality. As we have seen, in practice all taxonomic classifications will have fuzzy edges. Therefore, in terms of categorization, it is important to define categories such that Category X *excludes* Category Y (and Category Z, and Category A, and so on). In fact, unless one includes, as part of one's definitions, what the category excludes, one is not really defining anything and certainly not creating an adequate taxonomic distinction. As the philosopher Gilbert Ryle argued:

> In a country where there is a coinage, false coins can be manufactured and passed; and the counterfeiting might be so efficient that an ordinary citizen, unable to tell which were false and which were genuine coins, might become suspicious of the genuineness of any particular coin that he received. But however general his suspicions might be, there remains one proposition which he cannot entertain, the proposition, namely, that it is possible that *all* coins are counterfeits. For there *must* be an answer to the question: Counterfeits of what? [18; italics in original]

In a similar way, if we do not have an opposing distinction with which we can define something, then we are not really saying anything at all; we are merely saying, things are as they are or something equally inane. Good is defined in opposition to bad, white to black, pleasure to pain, fake to real, and so on.

In other words, in reality all taxonomic categories are analog, but for our pragmatic purposes as human beings we frequently have to impose a binary or digital structure on these analog distinctions. These are not necessarily digital, of course; the colors of the rainbow are generally divided in the West into an eightfold scheme. But, the principle that we frequently have to impose tight, nonfuzzy definitions on intrinsically fuzzy phenomena remains. It must always be remembered that this is done merely for pragmatic purposes; the resulting definitions have no privileged metaphysical status.

So, taxonomic classifications should be mutually exclusive. But, why should they be exhaustive?

2.5.3.2 Bucket Categories

Taxonomic classifications should be exhaustive for two reasons. First, we want to be able to categorize all the books in the library. It would not be much of a system if possibly large amounts of books simply had to be dumped on the shelves in no particular order.

More subtly, if there is a classification for books that cannot be classified, then there is always the possibility that this might become the biggest category. Say we have a book, *The Mystery of the Mind*. Under normal circumstances, we probably decide this should go under Psychology. However, if we now have an Other category, we might suddenly decide the following: "Hmm … well, it does not say it's about psychology, and the catalogue rules do not specifically say that books with mind in the title are to do with psychology; I had better put it in under Other just to be on the safe side." This can become habitual and lead to Other categories containing large numbers of cases that for all intents and purposes have not really been classified at all.

More specifically, loose categories that function *as if they were* an Other category (we call them bucket categories) can have a similar effect. So, a category of books that are not really about one thing specifically or books that touch on the subject of life in some form can again sweep everything up before them. We will look at specific examples of bucket categories in Chapter 3.

So far we have looked at MEE. But, we would go further and add that the best kind of taxonomy is a nested hierarchy. Why?

2.5.4 Nested Hierarchies

Again, let us take the example of psychology books and this time divide the super-category psychology into three subcategories. For the purpose of clarity, let us make them subdivisions of date, so that we emphasize the mutually exclusive nature of the categorizations. We have psychology divided into Psychology Books Written Before 1950, Psychology Books 1950–1975, and Psychology Books Written from 1976 to the Present. We can then subdivide these further into separate years and then into alphabetical categories representing first authors' surnames. Why have we done this?

The first answer is simply that this makes thing so much easier. Instead of running in and having to find the exact positioning immediately, the librarian can make the classification in stages. First, the librarian can find out whether it is to be filed under Psychology. Then, the librarian can look up the date for positioning at the subcategory level and the individual year level; and finally, it can be positioned in the correct alphabetical order.

Moreover, with a tight, nested hierarchy in which all the codes on the right-hand side or below the supercategory add up to the supercategory, we have a nice way of checking that there is no Other category, and that one is not needed.

However, there is a deeper point here. Imagine if all the codes (classification definitions) were laid out in list form, and we did not realize that psychology books written before 1950 were not hierarchically below (or nested within) the category Psychology. We might then generate spurious frequencies by double counting a book categorized under Psychology and under Psychology Books Written Before 1950. Arranging *and presenting* a taxonomy as a nested hierarchy is both logically sound and practically important in that it prevents this type of confusion. The model should be biological taxonomies, in which Life breaks down into Animal and Plant, and Animal breaks down into Mammal, Reptile, Amphibian, and so on. If one did not understand the nested nature of the hierarchy, that Animal subdivides into Mammal, Reptile, and Amphibian, we might think that Reptile was a distinction of the same kind (or level) as Animal and Plant. Naturally, in this effective taxonomy, there is no Other category.

2.5.4.1 Nested Hierarchies in SECAS

How does this work in a safety context? Let us return to our example from SECAS. We are going to be looking very much more closely at the use of individual codes in taxonomies in Chapter 3, but here, let us look at Figure 2.2 of the communication codes. These are nested, and hierarchical, and are to be read from left to right. Now, just imagine that all the codes were *flattened out*, such that the nesting was no longer apparent. Instead of something that looked like a flowchart, then we have what looks like just a list of codes. In no particular order, we would have Personal Communication, Information from Equipment/Plant, Content of Communication, Form of Communication, No Communication, Comprehensibility of Displays/Alarms, Attention to Displays/Alarms, Validity of Displays/Alarms, and Visibility/Audibility of Displays/Alarms. What now if we were simply told *that these were the codes we had to choose from*? Of course, now we know, because we have seen them in their correct order, that these codes, *presented like this*, are a mess. Put that out of your mind for a second and imagine that you are seeing these codes for the first time. Would the problems that you were going to face immediately occur to you? Let us assume for the moment that you are a harassed safety manager having to classify 30 documents in a hurry before lunch, that you have had minimal training, that the reports land on your desk, and that you are simply left to get on with it (not an implausible scenario, in our experience). And, let us say you stumble on a piece of text that indicates a problem with the audibility of an alarm. So, you pick Audibility. But, then you see this other code, Information from Equipment/Plant, and you think, "Well, actually, yes, it's an information problem." You either change your mind or pick that code as well.

What is the problem here? The problem comes with failing to realize that the first code is a subcategory of the second. You can conceivably pick both. Equally, you might pick the first or the second. Now imagine that process recurring. With a hierarchy you should have clear options at each stage, and the process of proceeding from major category to subcategory decisions should be clear.

So, what is the outcome of presenting codes in list form that are not MEE? The problem is that this increases the probability of *unreliability*. The more plausible options there are available, the greater the possibility of unreliable coding is.

There is also another problem, not in data input, but in data output. Again, imagine a taxonomy presented as if all of these were discrete codes. Output from the database would be virtually impossible to interpret as we would not know whether any data point represented an event that was also coded somewhere else in the list.

Therefore, as in the library example, we can see that a nested, hierarchical taxonomy with MEE categories is going to lead to the highest element of taxonomic reliability (consensus). The new SECAS codes, arranged in such a way, were subjected to a reliability trial; the results are shown in Tables 2.2 and 2.3. This was a two-day trial in which there were five coders on the first day and four on the second. Coders were given randomly selected event reports that coders had not seen before and were asked to code them using SECAS. Reliability was calculated, as before, by raw consensus figures between coders. As can be seen, the average degree of consensus went up to 61% (over the two trials).

There was a second element to the trial, which we discuss later. However, the key point is that we have shown that greatly increased reliability can be demonstrated with an MEE, nested, hierarchical taxonomy that coders follow through like a logic tree.

2.5.4.2 The Content of the Database

It is vitally important to remember at this point that the actual content of the data to be taxonomically arranged is irrelevant, provided it is of sufficient quality (not too much missing data, legible, audible, organized, etc.). For example, the SECAS data were sometimes termed a *minor event database* and came from a nonconfidential reporting system (as noted above). However, we have also worked on numerous other projects, one of which was CIRAS (Confidential Incident Reporting and

TABLE 2.2

Interrater Reliability Data for Strathclyde Event Coding and Analysis System (SECAS) (day 1; $n = 12$ events; 5 coders)

Paired Comparison	Coders 1 and 2	Coders 1 and 3	Coders 1 and 4	Coders 1 and 5	Coders 2 and 3	Coders 2 and 4	Coders 2 and 5	Coders 3 and 4	Coders 3 and 5	Coders 4 and 5	Average
Average raw agreement (Po) for 12 events	56%	51%	59.5%	57%	56%	68.5%	49.5%	59%	46.5%	61%	56%

TABLE 2.3
Interrater Reliability Data for SECAS (day 2; n = 12 events; 4 coders)

Paired Comparison	Coders 1 and 2	Coders 1 and 3	Coders 1 and 4	Coders 2 and 3	Coders 2 and 4	Coders 3 and 4	Average
Average raw agreement (Po) for 12 events	64.5%	59.5%	69.5%	63%	64%	77%	66%

Analysis System), which is a confidential system used in the U.K. railway industry. We have discussed CIRAS extensively elsewhere, and we are not going to dwell on it at any great length here [19].

However, briefly, CIRAS was a confidential system for gathering and collating safety-related data throughout the U.K. railway system. After an initial approach by the reporter, an experienced interviewer would carry out an interview (either on the phone or in person), which would then be transcribed and sent to the CIRAS Core at the University of Strathclyde. Here, the data were coded, and the resulting patterns were analyzed.

The methodology of data collection was therefore very different from SECAS, but the basic principles were the same. That is, there were data in some form that had to be gathered and analyzed in the form of a taxonomy. However, because it was a different project, a different taxonomy was necessary. The human factors aspect of this taxonomy was inferred from a 3×3 matrix (as with SECAS), which can be seen in Table 2.4. This time, the Front Line, Supervisory, and Managerial aspect was cross-referenced with the divisions Task, Procedural, and Communications; then, individual human factors codes were derived from this arrangement.

TABLE 2.4
The CIRAS (Confidential Incident Reporting and Analysis System) Matrix

	Job	Communications	Rules
Managerial	Staffing Resources	Communications Managers–Staff Communications between Managers	Rule Violations Procedures/ Documents
Supervisory	Task Management Training	Communication between Supervisors Communication Supervisors–Staff Communication Supervisors–Management	Tolerance of Procedural Drift Rule Violation
Front Line	Attention Fatigue Slips/Trips	Communication between Staff Communication Staff–Supervisors	Rule Violation

The whole system is shown in Figure 2.3 (codes were derived from all the elements shown in this figure).

The nested hierarchies can be illustrated by showing an example. Figure 2.4 shows the logic path to a single code, Communication Supervisor–Staff (to be applied, for example, when a communication from a supervisor to a work group has gone wrong). We can see from Figure 2.4 that a number of gates are gone through to reach this code, with MEE categories to choose from at each level. For example, Supervisory codes (see Table 2.4 and Figure 2.3) break down into Communications, Job, and Rules codes. The Supervisory Communication codes then break down further into Supervisors–Managers, between Supervisors, and Supervisors–Staff (the one chosen in this case). Each level shows a good example of MEE codes nested within the next level up in the hierarchy.

To illustrate this further, a screen shot of the electronic process (Figure 2.5) will show that, as in SECAS, coders were asked to move from left to right through the system until they ended up with a particular code. In this example (based on a software front end that we built ourselves), part of the text has been highlighted and coded as a Slip/Trip. (Slip/Trip can be seen as a code in the original matrix, under Job and Front Line.) In this software package, it is possible to pick out examples of selected codes and refer at the click of a mouse directly to the original report or text on which the code is based.

Yet again, the acid test of CIRAS was reliability. In a test of reliability/consensus, we provided overall reliability between three trained coders of 71.3%, which is above the acceptable level (see Table 2.5).* We deal further with reliability trials in Chapter 3, but the key point is not only that a system of MEE nested hierarchical categories can provide reliable data, but also that we would argue that the use of MEE nested hierarchies is the *only* way to guarantee reliability in coding. Remember, the CIRAS taxonomy and the SECAS taxonomy were different because taxonomies are situation specific. There is no one taxonomy for all situations. These, in a nutshell, are the basic axioms of what we call taxonomy theory.

However, these principles go far beyond this.

2.6 APPLICATIONS OF TAXONOMY THEORY

It cannot be stressed enough that, from the point of view of running a database (*not* necessarily an electronic database, but any form of storing data), we always simply have information to be collated and analyzed. The use of such taxonomic distinctions as qualitative or quantitative or any other taxonomic distinction one chooses does not change the task: data still have to be observed, collated, organized, and analyzed.

For example, consider video data. Running simulations of accident scenarios (e.g., in nuclear power plant simulators) or video-recording staff in a normal work environment is a simple way of obtaining precise data regarding how people actually

---

* Each cell of reliability data in Table 2.5 involves comparisons between two coders assigning multiple codes to each event for technical, human, and environmental factors, as well as possible or actual consequences.

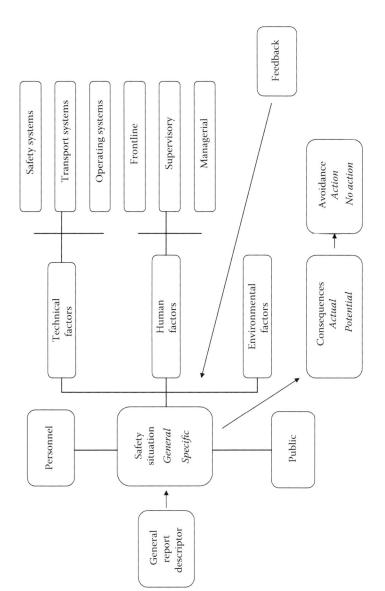

FIGURE 2.3 CIRAS (Confidential Incident Reporting and Analysis System) logic diagram.

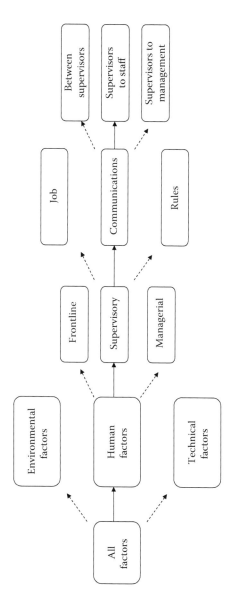

FIGURE 2.4 Logic gates for code Communication from Supervisor to Staff.

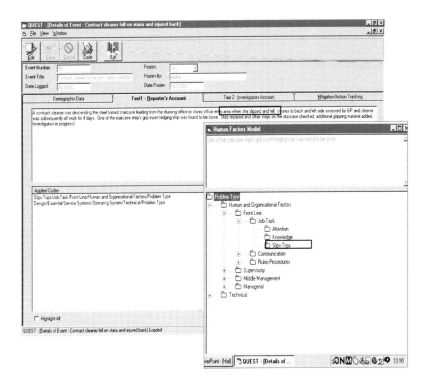

FIGURE 2.5 Screen shot from custom software showing logic for Slips/Trips.

behave in emergency/ordinary situations. This method of gathering data has been widely used in other fields (e.g., education) [20].

In these studies, situations (e.g., classroom situations) are videotaped and then coded by experienced coders. There is no reason this could not be done in a safety (or any) context; in fact, these methodologies have begun to be used in the field [21]. However, it must never be forgotten that the same principles are in use: one

TABLE 2.5
Interrater Reliability Data for CIRAS (Confidential Incident Reporting and Analysis System) (*n* = 10 events; 3 coders)

Event		1	2	3	4	5	6	7	8	9	10	Average
Paired comparison (%)	Coders 1 and 2	100	78.6	77.2	87.5	55.6	87.5	66.7	75	73.3	68.8	76.50
	Coders 1 and 3	69.2	38.5	63.6	70	55.6	100	80	53.8	78.6	64.3	67.38
	Coders 2 and 3	69.2	35.7	60	77.8	100	87.5	68.4	54.5	78.6	71.4	71.31

builds a taxonomy of classifications, which is then judged reliable or not (i.e., there is consensus in application or not). The basic rules (nested hierarchies, MEE codes) are the same.

It should be noted again that in a sense the videotaping of the situation may be superfluous. For example, in ethnography, raw (animal) behavior is observed by multiple observers, a coding hierarchy is built up, and then reliability is calculated. This is precisely the same process that we have discussed, a nested hierarchy and taxonomy is built up, observations are made, and then data are collated.

What about focus group data? Again, these seem different because we are used to looking at focus groups in a different way from interviews, but as we show in Chapter 3, with adequate reliability and a useful taxonomy, the amount of data we can produce from these databases (which is what they are) is greatly increased.

Again, the crucial aspect is always reliability (consensus), and we would argue this in a very strong sense. Are we arguing that data from a taxonomy that has not had a reliability trial should be considered dubious? Unfortunately (for some), that is precisely what we are arguing. We think that the burden of proof lies on those systems without reliability data, and that therefore systems should be considered guilty until proven innocent (as it were). People should assume that a system has poor reliability (and that therefore the methodology should be considered unproven, indeed *unscientific*) until reliability data are published.

2.6.1 QUESTIONNAIRE DATA

Although it is beyond the scope of this text to outline in detail questionnaire design and reliability issues, we would like to contend simply that data gathered through questionnaires should be considered in exactly the same fashion.* Taxonomies underpin *all* forms of questionnaire items. Forced-choice responses to questions designed to determine gender, occupation, place of work, and so on are perhaps easiest to recognize as taxonomic as they will involve discrete (exclusive) categories such as Gender breaking down into Male/Female. Yes/no/don't know questions similarly rely on the taxonomic properties of the choices, as do nominal (categorical) scales like choosing among strongly agree, agree, unsure, disagree, and strongly disagree. (It should be clear that these categories are designed as mutually exclusive and exhaustive, and that they are nested under the supercategory Degrees of Agreement.)

Often, the issue with questions like this is one of intrarater reliability, which refers to the consistency of responses from an individual on repeated occasions. It would be no use to have a list of occupational codes that were overlapping so that sometimes I put myself down as an Academic, sometimes as a Researcher, sometimes as a Psychologist, sometimes as a Human Factors Manager, and so on. Once more, a clear hierarchy and MEE categories are required if data are to be trusted. Readers will be familiar with the feeling of reading fixed questionnaire choices and being

---

* Data from content analysis, in which exactly the same procedures take place, should be considered in this fashion. Some software packages for content analysis have built-in procedures for calculating reliability, as they should. We have not dealt with content analysis here as it is so infrequently used in safety management, but it brings home the point that taxonomy theory is a meta-theory.

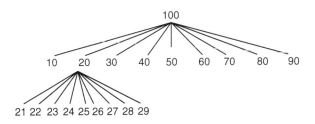

FIGURE 2.6 Numerical taxonomy.

unsure which to pick; this is *exactly* like a coder faced with choices to log an event and being unsure which choice applies.

It is also important to realize that questions requiring a *numerical* response (e.g., a request for age, length of experience, time into shift an accident occurred, temperatures, distances, etc.) have inbuilt taxonomies as well. The only difference is that we sometimes simply assume people can use numerical taxonomies reliably on different occasions (i.e., we do not do a test/retest study of people's coding of their age to ensure it is reliable, although maybe we should).

Figure 2.6 shows that numerical data can be considered taxonomic just as well as anything else, with the Hundreds (one hundred, two hundred, etc.) the supercategories (although, of course, you can go up as high as you want); these are subdivided and so forth. Sometimes, numerical codes like those for coding time intervals (seconds, hours, days, weeks) are collapsed to produce larger taxonomic categories, such as a five-point scale for frequency like: Every Day or More, Two to Six Times a Week, About Once a Week, About Once a Month, Never.

Like colors, there are many different ways of dividing numbers. For example, some people use parts of the body to count to 20 or 30 (and stop there). The Mayans used a base 20 counting system (as opposed to our base 10). The Pirahã tribe of Brazil has a system that simply divides things into one, two, and many [22].

This is an extremely important point because quantitative data are frequently considered somehow outside the scope of taxonomic thinking. Questionnaires then are often thought to be a way of generating objective data that are somehow ontologically different from qualitative data.

An example of how this is done would be to induce responses to a statement by asking people to, for example, agree or disagree with the statement by discriminating between a number (typically 5, 7, or 9) of numerical (or pseudonumerical) points. It is hoped that readers will understand by now we view this simply as a *taxonomy of agreement*. People are coding their agreement for the investigator. This is just like someone coding a video observation or a piece of text, and the reliability of the taxonomy is what matters.

2.6.2 THE SOLUTION TO THE PROBLEM

We have described in some detail how we created and tested a taxonomy to classify data in the nuclear industry. One aspect of the trial has not yet been discussed.

As well as the results reported in Tables 2.2 and 2.3, we tested the system by removing uncertainty regarding which aspects of reports to code. To do this, we asked one classifier to select aspects to code and then code them (as in the original trial). However, a second classifier was then instructed to code the same feature of the event that the first person had chosen. This eliminated unreliability in choosing what to code to isolate reliability in choosing a code to apply. It was found that, when this was done, reliability rose from 61% to 81%, clearly exceeding the necessary levels. What does this show? It shows that there are two distinct issues: choosing *what to code* from a field of text, report, investigation, or the like and *choosing which code to apply.*

It is simpler (and more reliable) to try to ensure that classifiers focus on similar and specific aspects (rather than allowing them to construct their own vague ideas regarding what they thought was important or to code in general terms what the whole text might have been about). In CIRAS, we ended up computerizing the whole process so that we could always relate codes applied to specific sections of reports (via a specially written computer program). We found this accountability forced classifiers to be much more rigorous, leading to similar salient aspects selected for coding. The philosopher Paul Ricoeur called this the judicial process of classifying a text [23].

However, it also has another implication, which solves the problem raised by Benner at the end of Chapter 1. This is because the rise in reliability shows that reports designed to feed the taxonomy would be easier to code than ones that left room for debate regarding what to code. Remember that Benner's problem was that there was no criterion for truth that was not specific to a specific accident analysis/ investigation methodology. Now, it may seem surprising to be returning to accident investigation theory, but we must remember taxonomy theory *by definition* applies to accident investigation because it applies to *everything*. The event reports in the database were in any case accident investigations by any definition.

What is argued here is that almost all (or perhaps all) currently used accident investigation systems (and minor event reporting systems, etc.) have been actually built backward. In other words, the methodology for gathering data was set up, then the database to order the data, and then the taxonomy to order *these* data. The debates discussed by Benner will rage interminably until one decides to build a reliable database with a reliable taxonomy *first*. Once this has been done, then one can go about building a method of data gathering that feeds the system.

This is a rather radical reversal of (so far as we know) everything that has ever been written about accident investigation, but it is logically impeccable. Certainly, one needs *some* raw data at the beginning (which might be a very small number of reports, discussions with process engineers, or observations of plant activity). As soon as information starts to come in, discussions should commence with the staff who have to use the database to create a workable and database-specific taxonomy (which may of course use lessons learned from other taxonomies, but the fundamentally unique aspects should never be ignored). Therefore, the way these things are normally done (i.e., with the data gathered in a more or less random fashion until it is decided that something ought to be done with them) is back to front. With a reliable taxonomy, you know that the data you are obtaining are useful, and you also know that they are reliable, coherent, and meaningful for the context in which they are used.

Then, reliability trials should take place; these are discussed in the next chapter. Once we have done that, then we can evolve a methodology to gather data to feed the system. In other words, once one has created the database with a working ontology (taxonomy), then one can create a data input system that feeds the system.

Benner's problem is thus solved. The accident investigation methodology that is best is quite simply the one that feeds the database, as now organized by our taxonomy. Our claim goes much further. We are not claiming that this is *a* solution to the problem. *We are claiming that this is the only possible solution to the problem.** Only by creating taxonomies that real people can use, which lead to real consensus in interpretations, will the subjectivity problem (i.e., the differing conceptions of truth across systems: Benner's problem) be solved. Until this is achieved, accident investigation, minor event reporting, and similar approaches will perpetually lack the rigor they crave.

2.7 CONCLUSION

To conclude, therefore, we have hopefully shown just how important taxonomic concepts are to our understanding of the world. Here, we make three presuppositions: taxonomies are created in some context; taxonomies have some sort of a pragmatic purpose; and a taxonomy is valueless if we cannot obtain social agreement on its use. What must be understood at this point is that *everything is taxonomic*, and that almost all of our debates in philosophy, politics, and so on are really debates about taxonomic classification.

There is a Charlie Brown strip in which Charlie Brown is beaten in an argument (yet again!) and goes to commiserate with Lucy. She replies that when questioned he should demand that the interrogator define his or her terms. The next time he is picked on is when someone comes up to him and says, "You're fat. Look at that stomach!" To which Charlie Brown replies, "Define stomach!"

This is a joke, of course, but it touches on an important truth. Most debates are actually about taxonomic categorizations, and most fights start from the assumption that words, sentences, and behaviors are not polysemous. But, they are. We have to define our use of words like *accident*, *cause*, *disaster*, and so on because the people we are arguing with may not actually share the meaning we have in mind. What we are attempting to do here is define a methodology by which we can measure, and then build up, the degree of social consensus that exists in taxonomic classifications. Unless we define our categories in opposition to other categories, progress cannot be made.

For example, if I describe a politician as a bad man, and you disagree, what is exemplified is precisely a difficulty in terms of achieving social consensus over a taxonomic classification. For example, we have a high degree of social consensus

---

* In fact, we would go further and argue that, for qualitative research in all subjects (management theory, social policy, sociology, etc.) with all methodologies (interviews, focus groups, questionnaires, etc.), reliability trials should be used. See the final chapter for more details.

over the figure of Hitler, whom most people would easily classify as evil. Other political figures have slightly lower degrees of consensus, and this is where the fights start.

In philosophy and science, many debates are in fact about the classification of events into taxonomic categories. Whether something is termed an accident or not, for example, is primarily a question of taxonomic classification, which in turn is a question of taxonomic *definitions* (in this case, what do you mean by the classification [word] accident?). Whether we decide to classify an event under the taxonomic classification accident will depend on what our definition of the classification accident is. This is not a simple matter, as we saw in Chapter 1.

Moreover, as we also saw in the last chapter, the use of the taxonomic distinction human error can be predicted in line with various biases in the classifiers. Classifications are not true in some abstract sense; they are functional and useful (and reliable) or not.

The central argument of taxonomy theory is that actions, words, numbers, or whatever are to be considered databases, stores of information that are used in certain contexts. For example, a book contains words and letters, which are primarily patterns of ink on a page (or, on a computer, pixels arranged in a certain order). We understand the book because we know the taxonomic classifications that enable us to read. We know that A is defined as Not B, Not C, Not D, and so on; furthermore, words are the same: Dog is defined as Not Cat, Not Zebra, and so on. If we did not have access to these socially created taxonomic classifications, then we would not be able to read. Therefore, the words themselves are understood and interpreted in precisely the same way as the books in the library. They are defined by their interrelationships, not (or at least not necessarily) by what lies behind them or to what they refer.

This should (finally) bring us back to technical rationality (TR) as opposed to our proposed exploratory methodology. The conventional (TR) approach to data attempts to look through the data, to the cognitive structures or real causes or laws that lie behind them. What we are attempting to do here is to look *at* the data. This is why it is important that we view all data in a consistent manner and not bother ourselves too much by wondering where they came from.*

2.7.1 TAXONOMY THEORY AND THE FUTURE OF SAFETY MANAGEMENT

We have therefore been left with a problem in terms of accident investigation; as we argue, as always this is a problem of taxonomic classification. The problem is this: what phenomena do we classify as causes? To repeat, *at the level* we are talking about, we are not actually on the front line kicking tin in the wreckage of an airplane

---

* We are doing this because, as a result of the arguments in Chapter 1, we are assuming that a TR approach will not work in this context, and that an exploratory approach therefore might. We might add that, in our experiences, such an approach does actually work (pragmatically), which is why we are writing this book.

or train.* Instead, we have texts in front of us, which may or may not be in electronic format, and we have to classify them in a pragmatically useful way. The way in which these will be classified remains dependent on the context in which we have to do the classification. There is not just ONE abstract or true classification of any event or situation: there are always choices.

We have stated that the main problem we had with the SECAS project, therefore, was the differences in classification (reliability). We must now make a statement that we do not have space to back up at the moment but to which we return again and again throughout this book: reliability is not *forced* consensus, in which (covertly) one person (the privileged classifier) makes the real distinctions and then uses influence (or whatever) to get everyone to go along with him or her. In our own reliability/consensus trials, we actually physically put the classifiers in separate rooms to prevent that from happening. We are more interested in the extent to which real *individuals* agree with one another; we are not trying to create some kind of "groupthink."

We make that point and move on to another: unfortunately, many people continue to overlook or ignore the fact that problems with reliability (consensus) quickly render a database effectively useless. To see why this is the case, in the next chapter we look at this issue a bit more deeply before looking at problems with some assumptions about probability in science and null hypothesis testing and the impact this can have on how *output* from a reliable database should be analyzed.

REFERENCES

1. Lucas, D.A., Organisational aspects of near miss reporting, in van der Schaaf, T.W., Lucas, D.A., and Hale, A.R. (Eds.), *Near Miss Reporting as a Safety Tool*, Butterworth-Heinemann, Oxford, U.K., 1991, p. 127.
2. Phillips, D., Jarvininen, J., and Phillips, R., A spike in fatal medication errors at the beginning of each month, *Pharmacotherapy: Official Journal of the American College of Clinical Pharmacy*, 25(1), 1, 2005.
3. Martin, P. and Bateson, P., *Measuring Behaviour: An Introductory Guide*, Cambridge University Press, Cambridge, U.K., 1993.
4. Borg, W. and Gall, M., *Educational Research*, Longman, London, 1980.
5. Brown, R., How shall a thing be called? *Psychological Review*, 65(1), 14, 1958.
6. Brown, R., How shall a thing be called? *Psychological Review*, 65(1), 16, 1958.
7. Rather, C., A sociohistorical critique of naturalistic theories of color perception, *Journal of Mind and Behaviour*, 10, 369, 1989.
8. Tarski, A., The concept of truth in the languages of the deductive sciences, in Corcoran, J. (Ed.), *Logic, Semantics, and Mathematics*, Oxford University Press, Oxford, U.K., 1956.
9. The main arguments against metaphysical realism are stated in Putnam, H., *Meaning and the Moral Sciences*, Routledge and Kegan Paul, London, 1978.
10. Sellars, W., Empiricism and the philosophy of mind, in *Science, Perception and Reality*, Routledge and Kegan Paul, London, 1963, p. 127.
11. Thompson, E., Color vision, evolution, and perceptual content, *Synthese*, 104, 1, 1995.

---

* This is because these people are following a methodology we have developed to feed the database.

12. Lakoff, G. and Johnson, M., *Metaphors We Live By*, University of Chicago Press, Chicago, 2003.

13. Sellars, W., Empiricism and the philosophy of mind, in Sellars, W., *Science, Perception, and Reality*, Routledge and Kegan Paul, London, 1963, pp. 127–196.

14. Kosko, B., *Fuzzy Logic*, Flamingo, London, 1993.

15. Levinson, S., Yélî Dnye and the theory of basic color terms, *Journal of Linguistic Anthropology*, 10(1), 1, 2000.

16. Priest, G., *Beyond the Limits of Thought*, Oxford University Press, London, 2003.

17. Robson, C., *Real World Research*, Blackwell, Oxford, U.K., 1993.

18. Ryle, G., Categories, in *Collected Papers, Volume 2: Collected Essays*, Barnes and Noble, New York, 1971.

19. Wallace, B., Ross, A., and Davies, J., Applied hermeneutics and qualitative safety data: The CIRAS project, *Human Relations*, 56(5), 587, 2003.

20. Stigler, J., Gonzales, P., Kawanaka, T., Knoll, S., and Serrano, A., *The TIMSS Videotape Classroom Study: Methods and Findings from an Exploratory Research Project on Eighth-Grade Mathematics Instruction in Germany, Japan, and the United States*, National Centre for Education Statistics, U.S. Department of Education, Washington, D.C., 1999.

21. Roth, E.M., Christian, C.K., Gustafson, M., Sheridan, T.B., Dwyer, K., Gandhi, T.K., Zinner, M.J., and Dierks, M.M., Using field observations as a tool for discovery: Analysing cognitive and collaborative demands in the operating room, *Cognition Technology and Work*, 6(3), 148, 2004.

22. Gordon, P., Numerical cognition without words: Evidence from Amazonia, *Science*, 306(5695), 496, 2004.

23. Davies, J., Ross, A., Wallace, B., and Wright, L., *Safety Management: A Qualitative Systems Approach*, Taylor & Francis, London, 2003, p. 85.

3 Taxonomic Consensus

3.1 RELIABILITY AND VALIDITY

If Chapter 2 was concerned with the idea that taxonomies lie at the core of safety management (and databases in general), this chapter and Chapter 4 are about working with such taxonomies. In fact, in the simplest of terms, this chapter is about *input to* databases, and Chapter 4 is about *output from* databases. To use the standard terminology, therefore, this chapter is about the *reliability* of taxonomies, and Chapter 4 is about the *validity* of taxonomies.

We need consensus on how we classify and arrange information, and we need to be able to do something meaningful with the structured information we obtain through reliable classification.

The order of Chapters 3 and 4 is no accident. Reliability comes first. We have stressed that what you get out of a safety system is only as good as what you put in. As many people in the field agree, "Reliability is a prerequisite for validity" [1,2]. We wish to discuss reliability first because we believe in the old maxim "garbage in, garbage out." We would take a hard-line stance on this and argue that an unreliable taxonomy (one with low consensus on discriminatory decisions) is not useful *in any way*.

For example, suppose we attempt to classify minor events in an industrial setting as *procedural* and *nonprocedural* (we would have to define these, of course). Now, it may well be that a safety manager would find it useful to have this information at hand, for example, when deciding how to spend a limited training budget. If people could not agree on what was to be placed in which category, then any decisions made on the basis of the distinction would be meaningless. One might as well assign the budget on a whim. This is why the first task is always to achieve reliability of classification. If a distinction cannot be made reliably, we cannot proceed to see if it might be useful because this possibility is immediately ruled out, and *we must find another distinction to make*.

It is sometimes not apparent to people why agreement is so important. Probably because of the dominant cognitivist paradigm in psychology (see Chapter 5), which views cognition as an individualistic internal process, the intelligence of the individual expert (and the view that a single expert account can and should have a privileged nature) has long been conventional in organizational circles. From such a viewpoint, one expert could simply give an authoritative account, for example, of what the procedural and nonprocedural problems were. However, Surowiecki stressed the importance of social consensus: "In part because individual judgement is not accurate or consistent enough, cognitive diversity is essential to good decision making" [3].

We think that taxonomic decisions should always involve multiple coders, among which comparisons can be made. However, there is an important qualifier.

People must be free to make their own decisions. The problem of groupthink leads to bad decisions by groups of people (we return to this concept in Chapter 7, in which we discuss the space shuttle accidents in some detail). Groupthink tends to occur when certain individuals become authoritative, dissent is minimized, and a group decision is agreed on. *This is not the type of agreement we are talking about here.* In fact, it is the opposite of what we are discussing. Hence, when carrying out a reliability trial, we make sure that coders cannot communicate during the trial (see the appendix); this is to make sure that coders are forced to make their own decisions. We show how to compare coders' decisions later in the chapter, but for now it should merely be made clear that, at the point of decision, individual judgments are required.

The *group agreement* is in essence a mathematical concept relating to the amount of overlap in individual judgments and is measured independently. So, we get a few people to make their own judgments, then look at how much they converge (or diverge). This avoids what we might call *pseudoagreement*, by which one person, usually someone in a position of power, makes a judgment, and everyone else says "Okay, you're right, boss!" We are talking about agreement between people who have their own opinions on events, causes, and consequences. The agreed answer in essence is the modal one, that is, the one most people have chosen, a property of the shared cognition of the group. Once more, Suroweicki argued cogently for this approach:

> This does not mean that well-informed ... analysts are of no use in making good decisions. ... It does mean that however well-informed and sophisticated an expert is, his advice and predictions should be pooled with those of others to get the most out of him. ... And it means that attempting to "chase the expert," looking for the one man who will have the answers to an organization's problems, is a waste of time. We know that the group's decision will consistently be better than most of the people in the group. [4]

We discuss expert judgment in Chapter 4 on probabilistic safety assessment and return to this matter in some detail in Chapter 8.

Before discussing group (social) consensus in detail, it must be noted that it is undoubtedly true that such reliability does not *predetermine* validity. The modal response to a question of discrimination or classification may not always be practically significant. For example, we could always reliably classify All Major Aircraft Accidents into a taxonomy with two categories: Those Involving Mount Erebus on Ross Island, Antarctica and All Other Major Aircraft Accidents. This taxonomy passes the input tests of mutual exclusivity and exhaustivity (MEE) described in Chapter 2. It is exhaustive for the purpose for which it is designed because the individual categories are nested within the higher level single classifier All Major Aircraft Accidents. In addition, the categories appear to be mutually exclusive (we can assume the second is defined such that it excludes the first, and we can reliably tell them apart).

So far, so good. But, we would hope it is obvious that a database holding this classified information would not be very helpful. The pragmatic usefulness is limited (almost zero) because of the lack of discrimination involved. We have one bucket category and one category that is almost wholly redundant (used once and unlikely, God willing, ever to be used again). This example is a little extreme, but in fact bucket categories or redundant codes are more prevalent in safety databases than one might think, especially if the taxonomy has evolved in an ad hoc manner.

In Chapter 4, we address pragmatic usefulness in some detail. Ultimately, we need both reliability *and* useful discrimination. But, we cannot have useful discrimination without reliability, so reliability is where we start (the Mount Erebus example simply shows we *can* have it the other way around: reliability without discrimination). Let us begin our two chapters on working with taxonomy by looking at consensus in coding safety events, with reference to real-life examples from safety databases. Two practical issues concern us: how to recognize where consensus might initially be a problem and how to quantify it via rigorous trials.

3.2 THE LOGIC OF TAXONOMIC CONSENSUS

The purpose of this section is to show how to begin to evaluate taxonomies in logical terms. The ultimate test of the reliability of taxonomy should be a trial, leading to a quantification of interrater consensus [5–7]. (It can also be useful to track individual coders over time for *intra*rater reliability.) Such trials (see the appendix) involve time and effort. It would seem worthwhile, in the first instance, to design a taxonomy that *looked like it might be* reliable. It is surprisingly easy to find examples of safety taxonomies that do not meet this basic first requirement on the road to evaluation, what we might call the eyeball test of reliability potential.

3.2.1 HIERARCHICAL CONFOUNDS

Let us start by looking at a taxonomy that we encountered as part of work we did with the British nuclear industry (see Chapter 2). The codes were designed to be applied to two classes of event report: minor event reports, which are logged at individual plants and nuclear plant event reports (NUPERs), which are held on a database at the industry Central Feedback Unit. NUPER events are by definition more serious, and reporting of these to the central unit is required as part of Health and Safety Executive regulatory activity and conditions attached to site licenses [8,9].

Reliability of assignment of root cause codes from NUPERs in a trial we conducted was 42% [10]. Perhaps an examination of some of the specific codes shed light on why coders were unable to agree to a useful extent on which codes to apply. The root cause part of the database involved in the trial contains 18 major categories, including Verbal Communication, which in turn breaks down into 11 specific codes shown in Table 3.1.

Remember, the 11 codes listed in Table 3.1 exist at the same level in this coding hierarchy (under the higher level code Verbal Communication). *This means they should be mutually exclusive.*

To explain why this is the case, let us go back to the concept of nested hierarchies (introduced in Chapter 2), for which classification proceeds from one level to another by breaking down individual codes into subcodes, which together make up the original classification. For example, we could classify cars by manufacturer (higher

TABLE 3.1
Verbal Communication Codes from within the Nuclear Plant Event Report (NUPER) Nuclear Event Coding System

1	Shift Handover Inadequate
2	Prejob Briefing Inadequate/Not Performed
3	Message Misunderstood/Misinterpreted
4	Communications Equipment Inadequate or Not Available
5	Receiver Not Listening
6	Communications Incorrect/Inadequate
7	Internal Team Communication Inadequate
8	Interteam Communication Inadequate
9	Supervisor Not Notified of Problem
10	Unidentifiable
11	Other

level), then break down each manufacturer into specific models (lower level). At each level, we want to make reliable choices. We can reliably discriminate between makes and then models because each contains exclusive categories. *Crucially, we want to make our discriminations one step at a time.* Although make, model, or even size or color of car might be a good way of classifying cars, each classification should be done at a separate level in a hierarchy so we do not end up with a choice between options, such as Silver Cars, Big Cars, French Cars, and Citroen Cars, which could clearly overlap for a single case.

If we look again at the codes in Table 3.1, we can see problems with exclusiveness because there are hierarchical confounds. Codes 7 and 8 appear to be subsets of Code 6, that is, if Code 7 or 8 is applied, then Code 6 *must* also apply. If the interteam communication was inadequate, then by definition communications in general were inadequate. There is no way to achieve reliable classification with such a system.

We might also ask the following: what is the relationship between Code 3 (Message Misunderstood/Misinterpreted) and Code 6 (Communications Incorrect/ Inadequate)? Code 6 appears to encompass all situations to which Code 3 might apply. If a message is misunderstood, then this is an example of inadequate communication, is it not? In fact, given the title of the whole set of codes here (*Verbal* Communication), it might be argued that Code 6 is actually a broader category than the one of which it forms a subcategory.

Hierarchical confounds like these are problematic because coders have no logical framework with which to guide choices. As another example, Code 9 (Supervisor Not Notified of Problem) is quite specific. It cannot be discriminated from Code 6 because, in truth, it does not belong at the same level in a hierarchy. We need a system that begins generally and becomes more specific or at least keeps different aspects discrete, like one for cars that classifies manufacturer, then model, then color, then year.

The alternative, as seen in Table 3.1, is a conceptual minefield in which coders have too many plausible choices for a given incident. If a message passed on between two team members does not have the desired outcome, a coder wishing to log the event could pick Code 3, 6, or 7 easily before even considering if Code 5 applied. It is not that any given choice is wrong, simply that the range of overlapping options increases the chances of another coder not making these same choices.

In this way, hierarchical confounds create unreliability.

3.2.2 CONFOUNDS BETWEEN CODES

Let us continue with a taxonomy that was sent to us for review as it was proposed for use by two of the major statutory bodies involved with the U.K. rail industry. The taxonomy split into three supercategories at level one: Incident Type; Primary Cause; Influencing Factors. Within the Primary Cause branch, there were seven second-level headings (i.e., subcategories), one of which was Errors by Railway Employees. This then split into five further third-level headings, one of which was General Employee Actions. The ten individual codes within this box are shown in Table 3.2.

Let us begin by describing a hypothetical (but we hope not too outlandish) event on the railways:

> *The subcontractor threw his hard hat in the direction of the lookout to attract his attention. Unfortunately, it fell short, bounced off the railhead, and landed in the path of another contractor carrying some electrical cable. In trying to step wide of the hat, the other contractor fell and sprained his ankle.*

How then are we to code the subcontractor's action? This seems logically to be a Misuse of Tools/Equipment, more specifically an Incorrect Use of Personal Protective Equipment (PPE). Caused Injury to Another Person is applicable, and we suppose what really happened was that the subcontractor Created Obstruction/Hazard, and

---

TABLE 3.2
Codes for General Employee Actions within Errors by Railway Employees within Primary Causes in a Proposed Hierarchical Taxonomy for the U.K. Rail Industry

Vandalism/Malicious Action
Disregard Rules/Instructions
Failure to Warn Others
Horseplay
Created Obstruction/Hazard
Lack of Care
Misuse of Tools/Equipment
Caused Injury to Another Person
Incorrect Use of Personal Protective Equipment (PPE)
Injured Stretching/Natural Movement, etc.

---

that the lookout was Injured Stretching. It is unlikely that the master rule book explicitly says "Do not throw your hard hat at the lookout"; nevertheless, Disregard Rules/Instructions would be a possibility. With access to the state of mind of the protagonists, we might even be able to decide whether all or any of Malicious Action, Horseplay, or Lack of Care should apply. In other words, in the example of one plausible event, *all* the codes could reasonably be applied to that event.

There are two possible replies to this. The first one is: "You made up that event to make the taxonomy look silly!" Well, we would counter that this is an important task for the safety manager starting a new job or addressing a new database — think critically — and try to construct cases in which the taxonomy you are working with might fall apart. In any case, it is hoped the reader considered this event to be plausible before reading that all the codes could be applied: it is a bit much to state, after the case, that the event was not actually plausible merely because it makes a taxonomy look unworkable.

A more profound response to this point, perhaps, would be: "So what?" So what if all the codes could be applied? It is here that things get interesting. What we would argue is that (as the reader will probably have guessed), in practice (i.e., in a real-world coding situation), it would be very unlikely all the codes would in fact be applied. Only one or two would be. But, which ones? As we have shown, in this case all the codes could reasonably be applied. If the reader wishes to know why this matters, cast your mind back to the library example in the Chapter 2. What we are saying here is that this example is like having a book that could reasonably be classified under psychology, physics, biology, or (for that matter) domestic science or poetry (or even any combination of those, given that texts can sometimes be classified under more than one category). In a real-life classification scenario, with so many choices available, there is a higher probability of disagreement (i.e., of Coder A classifying the event [safety event, book, or anything else] in a different way from Coder B). In our example, Coder A may well classify the event under Malicious Action, and Coder B might classify it under Horseplay. Why not? These categories do not appear to be mutually exclusive. (Note that we do not actually know this until we carry out a trial, but a coding scheme like this makes this outcome more probable.)

In very simple terms, good taxonomies cannot be made to look foolish. Codes will be defined so that they overlap as little as possible and so that there is usually a good reason for choosing one rather than another. The proof will be in the reliability data (see Section 3.4), but if the taxonomy is plausible, most events will at least appear that they might be reliably classified via the system. Remember this is a list of codes within a primary cause taxonomy. Although no list of instruction for use was available to us (perhaps mercifully), we assume a single code was to be chosen for each event because this would usually be the case within a primary cause model. Does this look at first glance like something that a group of safety managers could do reliably? We think not. To think back to our principles outlined in Chapter 2, are the codes mutually exclusive? Can horseplay demonstrate a lack of care? One might actually ask, can Incorrect Use of Personal Protective Equipment (PPE) ever *not* be Misuse of Tools/Equipment? Or, would a malicious action ever *not* involve some sort of disregard for rules/instructions?

3.2.3 Confounds between Dimensions

It is also possible to look at a list of codes like this (we noted they are nested within a larger box called General Employee Actions) and note a further aspect, which we call Dimensional Confusion. Here, the question to ask oneself is, how are we to make distinctions between the categories? Look at the first category, Vandalism/ Malicious Actions. This code implies a volitional aspect in that the person had a particular (in this case destructive) motive for his or her behavior. Not all the codes are like this. Some are what we might call Behavioral, that is, descriptions of a type of action, for example, Creating an Obstruction, which says nothing of whether this was done intentionally.

Thinking about these codes for a second, it becomes clear that if one creates an obstruction *deliberately*, then the idea of vandalism suddenly becomes possible. In other words, we have a sort of epistemological confusion: the codes are not mutually exclusive. We have on the one hand a description of the event (an obstruction was created) and on the other a description of a cognitive internal state that allegedly caused the event. Not only do we now have two ways of describing the same event (never a good idea because it leads to unreliability), but also cognitive codes are intrinsically unreliable as they are supposed to be applied to conceptually unobservable events.

If you and I see a bear, we will have very high reliability regarding how it should be classified based on its outward form, behavior, or big hairy paws (i.e., we both shout, "It's a bear!"). On the other hand, if we are asked what the bear is thinking about, the consensus between us will drop considerably. It is the same with actions. In the final analysis, when a coder is faced with the question, "Was the action well intentioned or deliberate or not?" the coder may often have no idea, or no way of finding out, and may often be of the opinion that it does not matter very much, with the important aspect the creation of an obstruction that can have an impact on safety rather than the (alleged) cognitive cause.

Let us consider a final example that shows clearly what we mean by dimensional confusion. In this case, the taxonomy is designed to classify unwanted (adverse) events in the U.K. National Health Service. There are 20 supercategories, including General Complications of Treatment. This particular supercategory breaks down into 22 subcategories, including Wound Problem. The seven individual codes for Wound Problem are shown in Table 3.3. Some of these break down into further categories, indicating a taxonomic hierarchy. The original code labels are included to show the hierarchical organization.

One would have to be more familiar with medical terminology than we happen to be to say much about the ability of coders to distinguish among some of the categories. However, what is of interest here is the first pair of categories/codes (i.e., Swab Left in Wound and Instrument Left in Wound). Although these are not hierarchically organized, it is clear that *conceptually* they come under a (hypothetical) more abstract heading Something Left in Wound (we presume swabs and instruments cover most, if not all, things that might conceivably be left in a wound).

The question that we would ask is, what are these codes doing in a box called wound problem alongside codes such as Seroma (a seroma is fluid accumulation as

TABLE 3.3
Codes for Wound Problem within General Complications of Treatment

W0L0	Swab Left in Wound	
W0L1	Instrument Left in Wound	
W0L2	Wound Infection	
	W0L20	Wound Abscess
	W0L21	Cellulitis
	W0L22	Fasciitis
W0L3	Seroma	
W0L4	Wound Dehiscence	
	W0L40	Deep Wound Dehiscence
	W0L41	Superficial Dehiscence
W0L5	Failure of Wound Healing	
W0L6	Wound Skin Necrosis	
	W0L60	Minor Wound Necrosis Requiring Dressings
	W0L61	Major Wound Necrosis Requiring Surgery

a product of tissue inflammation)? Surely saying that Leaving Something in a Wound is a Wound Problem is like saying that Leaving Something in a Taxi is a Taxi Problem. Thinking of things that could happen in a hospital, it could be argued that, in taxonomic terms, Instrument Left in Wound is more like Bucket Left in Corridor (i.e., a rule violation) than it is like a Seroma.

This confusion of the behavioral and the medical is a bit like the confusion of the behavioral and the cognitive in the railway example. The implications are once more felt in terms of overlap and lack of reliability, for example, when swab left in wound leads to an infection or where infection means a wound does not heal properly. (However, there will also be the issue of obscure output, for which frequencies of Wound Problems will include such incongruous events as serious breaches of procedure [things being left in wounds] alongside minor medical matters [superficial wound dehiscence].)

The simplest way to avoid overlap (i.e., to strive for mutually exclusive codes) and to provide meaningful output (see Chapter 4) is to try to avoid confusing different dimensions for distinguishing among codes *at a particular level* within a coding hierarchy. We noted in Chapter 2 that it is perfectly acceptable to switch dimensions as we go up or down a level, for example, to classify books in a library in terms of their subject at one level, then within each subject to code via the alphabet, and within letter of the alphabet to code via date of publication, and so on.

3.2.4 EXAMPLE OF A HIERARCHICAL TAXONOMY WITH DIMENSIONAL LOGIC

We think it useful at this point to include an example of a reliable coding frame with proven pragmatic usefulness to make this point clearer. Figure 3.1 shows a small set of codes from within a hierarchical taxonomy we have employed in a

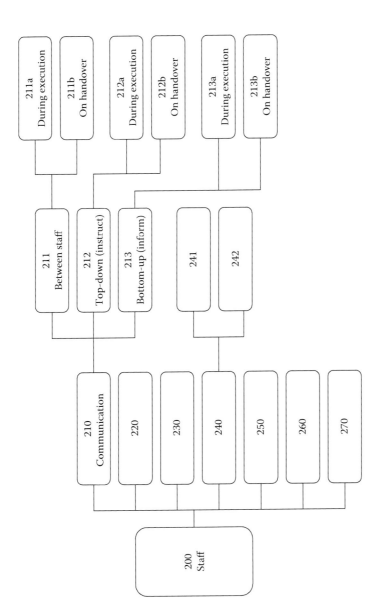

FIGURE 3.1 Communication codes within staff codes within the error-promoting conditions (EPC) branch of a hierarchical taxonomy.

high-consequence operational setting. The codes shown are codes for Communication Issues within a set of codes for human (staff-related) factors.

The highest level code here is Code 200 (Staff). At the next level, the distinction is between types of staff problems (the codes with a zero on the end). If we look at the top code 210 (Staff: Communication), we can see it breaks down further. At the next level, the codes 211, 212, and 213 are distinguished from one another in terms of the direction of communication within the organization (e.g., Code 211 denotes Communication between Staff). Finally, these break down further (the codes ending in *a* or *b*) in terms of whether the problem occurred during a shift or on shift handover from one work group to another. So, one of our final codes, Code 212b, denotes Staff: Communication: Top-Down: On Handover.

This is what we mean by different dimensional distinctions being kept apart. We first decide what the problem is, and in the case of the communication branch, code first the direction of the communication and then finally whether it took place on handover or not. We have found it useful in improving reliability to make distinctions in incremental steps.

As indicated by the shape of the branches in Figure 3.1, not all of the codes may break down to the same level. The important thing to keep in mind is that, when producing frequency counts or more complicated output (see Chapter 4), like should be compared with like, that is, comparisons should be among codes at a particular level at which a particular dimension can be studied.

We hope such examination of specific aspects of safety databases gives a flavor of the work with taxonomies that can be profitable in advance of reliability trials, which will ultimately show how well they function. However, there is an important point to be made. We might expect some readers to note how *presumptuous* we are about reliability. We have been talking about how some taxonomies look unreliable (i.e., we perceive people may have difficulty assigning safety events to the choices available to them). You might ask whether taxonomies should be given the benefit of the doubt. Should codes not be innocent of unreliability until proven guilty? Would this not be more *scientific* than simply eyeballing the schemes and deciding to change them around?

3.2.5 DEGREES OF BELIEF

There are two answers to the above question: first, we would wholeheartedly agree that reliability tests alone show the discrimination potential of a taxonomy. However, in practical terms, it would seem unusual to test a system that one found confusing rather than one in which the design had some thought put into it before testing (this would be the principle, one could argue, behind the preoperational testing of *any-thing*). Second, and more fundamentally, objections to hunches about probability (e.g., the probability of people agreeing on coding) usually arise out of a particular interpretation of probability that does not allow for this subjective guesswork.

Let us take a simple example of two coders (Alan and Beryl) and two codes (X and Y). Let us ask a simple question: what are the chances of them agreeing on the classification of a sample of events using the two codes? In some of the examples above, we *guessed* that there would be disagreement after examining the coding definitions, such as if we felt there was overlap in the codes or confusion in dimensions coded. So, if the two coding choices were Malicious Action and Creating Hazard, we might assume that there would not always be agreement because the two are not mutually exclusive, and one is based on motive, the other entirely on behavior. Although this estimate is based on looking at the coding choice, we might also make a guess based on what we know about Alan and Beryl. If Alan has a psychology background and Beryl has an operational one, we might expect different preferences, with Alan using the motivational code more often.

If you think it is plausible to consider codes and coders in this way and estimate agreement, you are accepting a view of probability that defines it as the degree of belief someone has that something will happen. This is known as the Bayesian view of probability.

Our adoption of a (broadly) Bayesian position on probability is why we feel no paradox in making statements about taxonomies *prior* to testing codes for exclusiveness. With a Bayesian perspective, it is perfectly acceptable to estimate probabilities of agreement in advance and revise these once the first reliability trial has been carried out. In fact, this process (using prior estimations) is at the heart of Bayesian inference.

However, it should be noted that there is also another view of probability, what is usually termed the frequentist or objectivist view. The English statistician and geneticist R. A. Fisher (1890–1962), whose work still dominates much of applied statistics, was particularly influential in developing frequentist statistics. Fisher would have disagreed (vehemently) with the view we stated regarding that it may sometimes be reasonable to guess the odds of agreement (or anything else) based on our experience.

Let us ask the frequentist the following: what are the chances of Alan and Beryl agreeing when using our two codes? The frequentist answers this question by first stating some things to be the case. First, the true value of this probability is unknown or unknowable (you cannot make poor Alan and Beryl code events forever to obtain the true or ultimate answer). Second, you are not allowed to guess until you do a study or trial to answer the question (so no matter what the codes are and who the coders are you can neither assume prior knowledge nor assign a prior probability of agreement). Finally, even when you carry out a trial and observe some data, you cannot use these data to make guesses about what will happen in a *subsequent* trial. Each new trial must start from the same position of ignorance as the first one; that is, you cannot estimate (infer) that the agreement this month between Alan and Beryl will be about 60% because that is what it was last month (what you *can* infer from the frequentist position we outline in a moment).

This is not a book about statistics per se, but the type of science described in Chapters 1 and 2 is compatible only with a subjective position on probability, so we now take a moment to outline clearly the difference between the Bayesian position and this traditional frequentist approach that, it must be said, lies at the heart of most approaches to risk estimation, reliability testing, and indeed to scientific experimentation.

3.3 APPROACHES TO PROBABILITY

Most of the procedures commonly used to make statistical estimates were developed by statisticians using a certain concept of probability, namely that the probability of an event has a true (objective) value which we cannot know, so we have to generate a probability based on the concept of a large (or infinite) number of repeated trials. Quantitative assessments for industries (for example, the probability of a particular technical breakdown or a particular human failure under a standard Human Reliability Assessment; see Chapter 4) are almost all framed in these terms.

Perhaps an example would help. Let us imagine you have a new piece of equipment, say a valve in a pipe. You want to know how reliable it is, that is, the proportion of times it will break down within the number of times you use it. Quite sensibly, you put it into operation and run the process for which it was designed (say, a sluicing process) a number of times. Then, you count how many times it fails. For example, you may run it 1000 times and find three failures.

3.3.1 PRIOR EXPECTATIONS

The main problem for the frequentist comes when predicting what will happen over the *next* 1000 trials. The Bayesian safety engineer can now state that he or she *expects* the valve to be reliable at a level of 3/1000 (0.003), but the frequentist cannot. The frequentist is not allowed to make such an estimate about what will happen next (called a prior estimate).

As we stated in the example of coding trials, the frequentist has to assume (or pretend to have) no prior knowledge at the start of each new trial. This idea of the independence of trials is important. Frequentist probabilities are based on the idea of many independent trials. Remember, the frequentist assumes a real (sometimes called absolute or ultimate) probability that is unknown, or unknowable, and can only be estimated from data.

Frequentists often give the example of coin tossing, with each toss a trial. If we toss a coin and it comes out heads four times, we have no reason (within a frequentist paradigm) to expect the next toss to be a head or for heads to come up more overall. The observed relative frequency of heads and tails in this case (from a small and finite sample) is *irrelevant* in terms of the final probability. In other words, as we are always told, even if you toss a coin 200 times and it comes up heads, the odds of heads on the next toss also are 50/50 because, for the frequentist, the probability of a head or a tail has nothing to do with what you think is going to happen next.*

The coin example can be misleading. The frequentist interpretation (that each new trial is independent of what has gone on) sounds entirely sensible in this context because people intuitively think of coins as unbiased (that is, the numbers of heads and tails will end up roughly equal). So, original sequences of four heads or four

---

* Although their discussion is beyond the scope of this text, there are different types of frequentism. The classical interpretation of probability is structurally similar to finite frequentism in that probability is underpinned by a ratio of cases. The difference is that finite frequentists consider actual outcomes of repeated trials and classical theorists consider possible outcomes of a single experiment or trial. The different stances lead to similar problems, including that of reference class (see Section 3.3.2).

tails do not lead to different expectations regarding the next toss. It is still a 50/50 call. We are all frequentists at heart when it comes to coin tossing, but the question is, how good is the coin-tossing analogy for real life?

Suppose we have a traction unit, reactor, circuit, or any other type of appliance or machine, and it keeps on failing. Do we continue to assume a position of ignorance and approach each new situation afresh? Or, is it more sensible to expect it to fail again? Do we not go out and buy a new one? To repeat the example above, if two coders have a 60% raw agreement in one trial, then do we not start the next trial with a rough idea that the agreement will be close to this value? Is this expectation not *precisely* what we do with probability every day? In fact, is this not a very important part of the job of the safety manager: to manage (in the short and medium term) based on (a) experience and (b) current knowledge of the state of the system?

This is the point at which Bayesian approaches are different. If we start with a run of four heads, then this is allowed to influence our expectation of a head resulting from the toss. If you like, from then on we imagine the coin to be biased until proven otherwise. Less abstractly, a piece of kit that breaks down would be thought of as unreliable until such time as it started to work more often in the manner its manufacturers told us it would when we bought it. The valve in the above example would be thought of as fairly reliable (3/1000 failures) until data showed otherwise. Or, in the context of the reliability data discussed later in this chapter, a taxonomy that people could not use to code events in the same way would be thought of as unreliable until redesigned and retested.

In a Bayesian view, we never approach things in a presuppositionless way. We always have some guess, view, or hypothesis regarding how something might or might not work. So, "the prior summarizes what we knew … before the data became available, the likelihood conveys the impact of the data, and the posterior contains the complete information available for further inference and prediction" [11].

In other words, say you buy a washing machine, personal computer, or whatever and someone asks you, before you unpack it, "Will it work?" From the frequentist point of view, the answer is "I haven't the faintest idea!" In reality, we are likely to have read a review of this piece of equipment, to have worked with similar models, to use one at work, to know friends who have one (who might have recommended it to us), and so on. This social knowledge constitutes our prior (i.e., prior to the first trial) expectation of how it might work, and it is *not* irrational or random. What we do when we boot up our personal computer for the first time (or switch on our washing machine or whatever) is then to revise our prior and then continue to revise it the more we use the piece of equipment: this process never ends. For Bayesians, probability is inherently dynamic. There is no objective or absolute answer to the question, what is the probability of x? There is only a pragmatic answer that we can give at one particular point in time.

3.3.2 REFERENCE CLASS

Another problem with the frequentist assumption that there is a single fixed objective (ultimate) probability is that this must be based on ratios of events, such as the ratio of valve failures to valve processes in the example above. Definitions of events (failures, processes, pieces of equipment) can be fairly tight in engineering, but once we introduce people and start talking about sociotechnical systems, things get trickier.

For example, suppose I want to take a frequentist position and estimate my chances of getting home by bus tonight without a crash. For the frequentist, I will only take one bus journey home tonight (i.e., it is a single event), and I will either crash or will not crash, so the ultimate probabilities are 1 or 0. A meaningful frequentist probability is derived by assuming that other events are similar to the one in which I am interested. I need to compute a ratio based on numbers of more general bus journeys home and crash frequencies. But, it is hoped readers will have started to realize by now that these categories (like other categories applying to the real world and especially those involving humans) are not as tight as we would like; in fact, as we saw in Chapter 2, they are fuzzy.

What about bus minijourneys? Should they count? What about bus journeys within the bus depot? Should these events be seen as part of the category in question? Answers to these questions depend on what I actually want to know. Do I actually want to know the frequency of bus crashes that give probabilities that only apply to me? Or, do I want more general probabilities, that is, probabilities that apply not only to me, but also to you (and other bus users)? If so, I might want to include all bus journeys taken by every human in the world (over a given time period). If these data were not available to me (as seems likely) or if I was concerned about more localized issues, would I perhaps limit myself to statistics relating to European buses, just U.K. buses, Scottish buses, or Glasgow buses or just the number 42a to Drumchapel on a Wednesday night in the rain?

Then, even when we have defined bus journey and gotten our data (from whatever source), we still have to define crash. Should minor scrapes count as crashes or just ones with passenger injuries (or driver injuries, and how do we define injury?)? What about when the bus was undamaged but a car had to swerve and crashed into another vehicle? Was that a bus crash? What about a crash between two buses? Is that one event or two?

These are questions relating to what is called our frame of reference for calculating the supposedly objective risk of me having a crash. The *subjectivity* (range of possible choices) inherent in choosing the reference classes for journeys and crashes is a problem for the objectivist position on probability. Simply, there are always *choices* that are made by real human beings when we create the categories (taxonomies) from which frequency data for frequentist ratios are established and probabilities are derived.

To admit this is to blow a major hole in the argument for frequentist probability because frequentists argue they derive objective probabilities. The objectivity is, unfortunately, secondary. The process is underpinned by the creation of categories (like Crash, Bus, and so on). Of course, these categories are taxonomic categories. They are useful and agreed on, but they do not have any privileged metaphysical status. Therefore, the idea that frequentist probabilities are objective in the sense that one single true probability can be obtained is a myth. Instead, frequentists smuggle in their subjectivity via the back door, all the time denying that this is what they are doing.

Now, at this point something has to be conceded about Bayesian probabilities: they are also subjective. However, the subjective element to calculating probability is faced up to, not brushed under the carpet. We would say that this is a good thing, but it could be argued that, for the Bayesian, taxonomic classifications are no easier to define and apply reliably than for the frequentist.

We outlined a methodology in Chapter 2 that we hope goes some way toward dealing with such problems. Deciding which categories to use in determining probabilities (framing) we would see as a taxonomic issue within the bounds of taxonomy theory and a working task for the taxonomic engineer. It cannot be problematic unless we take the (misguided) view that some real probability (usually called an absolute or an ultimate probability) is what we are trying to discover.

If we translate this into our language of taxonomy theory, we can see that *probabilities* are the ratio of one taxonomic category to another, no more and no less. Their *objectivity* is simply the extent to which we can agree on which events should apply. Once more, the central issue becomes that of taxonomic reliability. Within a given context, what is important is that people can agree on the defined events and the classes to which they belong, that is, the numerators and the denominators in the probabilistic calculations. Once we have made this arbitrary (i.e., subjective), but hopefully pragmatic, decision, the working out of the numbers can be called objective if it makes people feel better.

In practice, the frequentist's inability to give statements of expectation or belief often leads to assumptions of randomness or chance prior to data observation. So, the starting position for a choice between two codes will always assume a 0.5 chance of choosing each regardless of what they look like.

The Bayesian view of probability, on the other hand, is related to degree of belief. It is a measure of the plausibility of an event given an acknowledgment of incomplete knowledge. This is very different. *Bayesians say only the data are real, not the hypothetical absolute probability.* Bayesians view this absolute probability as an abstraction and as such take it that some values are more believable than others based on the data and their prior beliefs. The Bayesian thus constructs what is called a credible interval tempered by prior beliefs. Then, the Bayesian can say things that the frequentist cannot, like "There is a 95% chance this valve will be reliable within a certain range."

Now, we concede that this is all a bit technical for nonstatisticians (like us). But, we should be able to convince the reader that Bayesian approaches have the ring of common sense. An interpretation of reliability data in line with Bayesian thinking would be as follows: "Before testing the system, I was 80% sure that two coders would code an event in the same way; now, after looking at the data from my study, I am only 68.5% sure that coding is reliable."

Importantly, what is not valid according to frequentism is the assignment of probabilities to statements of confidence (ideas), such as, "The proposition that the reliability of a coding scheme is between 70% and 80%." It might be noted, however, that in our opinion, human factors experts and safety scientists, despite common use of frequentist statistics, seem to talk naturally about the probability of such hypotheses (for which true frequentism has no terms). If, at a safety conference, someone is asked what the probability is that their accident rate is between 2 and 2.5 per million train miles, it is unlikely the reply will be, "The question is nonsensical because the accident rate is a constant of unknown value." The answer is more likely to involve a guess (based on previous experience), is it not?

We have adopted a Bayesian perspective here because it seems sensible. If we measure reliability, it seems that what we want to know about is the chance of people agreeing with one another, say, 70%–80% of the time based on our experiences.

Given that this is the case, we would argue that researchers should look more closely at Bayesian, rather than frequentist, methods for calculating the reliability of binary decisions. For example, Lee and Del Fabbro [12] proposed a Bayesian equation, the BK coefficient of agreement for binary decisions (i.e., what we would call reliability). Like all Bayesian approaches, BK uses observed data to revise prior beliefs. So, as a picture of agreement builds, changes in actual coding rather than ratios alter the calculation (i.e., the more actual agreements there are, the higher the coefficient becomes). Davies et al. [13] recommended consideration of the BK coefficient for correcting raw consensus, and we have not found it difficult to perform BK calculations. However, we should perhaps turn here to the very basic calculations for agreement before looking at a common frequentist approach and some alternatives.

3.4 QUANTIFYING TAXONOMIC CONSENSUS

We have appended (see the appendix) an outline of simple procedures for carrying out a trial of interrater coding reliability. The result of following the basic steps in the appendix will be that we have a set of codes assigned by different coders to the same set of events. We now want to discuss quantification of taxonomic agreement via examination of codes assigned during trials of this type.

It is usually best to decide a priori what level of agreement will be acceptable for the type of classificatory system of interest. For coding of event reports, for example, we feel the system is essentially observational, with written verbal (or linguistic) events observed and coded as they appear in reports or transcripts. Borg and Gall argued that in observational studies a level of 70% is agreed as acceptable [14]. Although this is entirely arbitrary, the fact is we do have to define success and failure before we start our trial; otherwise, we have no basis on which to decide what to do with the taxonomy afterward. This is yet another example of imposing a binary structure on an analog phenomenon (coding) and another demonstration of why, despite that this to a certain extent is an arbitrary process, it is also a necessary one.

3.4.1 RAW AGREEMENT

Let us examine in stages what calculations will be necessary to calculate the reliability of a coding taxonomy. Suppose we have a simple case with our two coders, Alan and Beryl, and our taxonomy has five mutually exclusive codes denoted A, B, C, D, and E. Suppose also that we have ten events to code into the system. A hypothetical assignation of codes is shown in Table 3.4.

Interrater agreement is first obtained via the index of concordance [15]. This is simply $a/a + d$ where a = agreement and d = disagreement (see rightmost column in Table 3.4). In this case, we have 8/10 = 0.8. Multiplying this by 100 gives a raw

TABLE 3.4
Codes Assigned by 2 Coders with 5
Choices to 10 Events

Event	Code Assigned		Agree (a)/Disagree (d)
	Alan	Beryl	
1	A	A	a
2	B	B	a
3	C	C	a
4	D	D	a
5	E	E	a
6	A	A	a
7	B	B	a
8	C	E	d
9	D	E	d
10	E	E	a

agreement of 80%. This is called raw because it has not been corrected, which is a concept we discuss in more detail below.

However, although most people would not argue with our calculation so far, one thing has already gone wrong. Raw agreement derived in this way is an omnibus agreement score, that is, a catchall measure that does not tell us which codes we have agreed on and which codes we have failed to agree on.

This problem aside, people usually have no difficulty calculating the index of concordance/raw agreement for two coders. Problems may arise when there are more than two coders and thus multiple possible coding comparisons.

The number of pairs in a group of coders is obtained from the formula $N(N - 1)/2$ where N = the number of coders. So, for three coders there are three pairs, for four coders six pairs, for five coders ten pairs, and so on. We have previously described how the coding of multiple coders can be misinterpreted [16], but the point is worth making again here. Suppose we have four coders, A, B, C, and D. The codes they assign to a single event are X, X, X, and Y, respectively. That is, only coder D has coded a Y, and three of the four codes are the same. The number of X codes applied can seduce the unwary taxonomic designer. Many reporters would claim a raw agreement of 75% in this case. We understand that the common sense logic behind this is intuitively appealing (three of four = 75% agreement), but we have to use the above formula to remind ourselves there are six pairs (i.e., possible agreements) with four coders: Coder A versus Coder B; Coder A versus Coder C; Coder A versus Coder D; Coder B versus Coder C; Coder B versus Coder D; and Coder C versus Coder D. Of these six possible agreements, in this case there are three disagreements (Coder A versus Coder D; Coder B versus Coder D; and Coder C versus Coder D). We know this is counterintuitive, but the actual agreement for this small sample of codes is 50% (three agreements for six pairs), not 75% (three codes the same out of four). This is below our general criterion for acceptability of 70 [17].

Taking care not to fall into this trap, we now have a raw agreement score that gives a basic measure of how reliable our taxonomy is.

3.4.2 Correlations and Reliability

A second brief point is that it is important that taxonomic consensus is established for individual events or cases. We have previously shown that confusion can arise during the testing of safety management systems because this principle is overlooked. If reliability of a coding frame is equated to consistency in *overall patterns* (usually measured via correlation), then misconceptions can arise when it is assumed without justification that people can apply the same categories to individual events [18,19].

It is fairly straightforward to prove this is the case. Suppose we look at frequencies of codes assigned during a reliability trial by two coders, Alan and Beryl, using a five-item taxonomy: A, B, C, D, and E. Let us assume they have coded a sample of 15 events or cases, and Alan has assigned the codes with respective frequencies for A, B, C, D, E of 5, 4, 3, 2, 1. Beryl has used them with respective frequencies of 5, 4, 3, 1, 2. Now, a quick comparison appears to support high reliability. In fact, the rank order correlation will come out at 0.9. To illustrate why this can be misleading, we have included in Table 3.5 a possible arrangement of individual codes from this imagined trial. Codes assigned by Alan are arranged in order for simplicity.

Remember the correlation here between the codes assigned is 0.9. But, if we add the raw agreement (in the final column), then we see it comes out to 5/15 = 0.33

TABLE 3.5
Codes Assigned by 2 Coders with 5 Choices to 15 Events

Event	Code Assigned		Agree (a)/Disagree (d)
	Alan	Beryl	
1	A	A	a
2	A	B	d
3	A	A	a
4	A	D	d
5	A	E	d
6	B	A	d
7	B	B	a
8	B	C	d
9	B	C	d
10	C	C	a
11	C	A	d
12	C	B	d
13	D	B	d
14	D	E	d
15	E	A	a

or 33%. We can see that two coders with a very similar overall spread of codes can fail to assign the same code to the same event with an acceptable degree of reliability. *Taxonomic reliability has to be calculated using comparisons between codes for individual events* (in this case, event reports) (see the appendix). What we want to know is that you and I will code the same event/report in the same way.

Let us assume we have not fallen into the trap of equating correlation with reliability and have calculated raw agreement for individual cases. At this point, most reliability trials involve proceeding beyond these simple calculations of raw agreement to some further statistical analysis. The most common coefficient of agreement used in such studies is undoubtedly Cohen's Kappa coefficient [20–22], to which we now turn briefly.

3.4.3 COHEN'S KAPPA

Kappa is a simple formula for correcting the proportion of agreement achieved by independent judges for the number of agreements that would be expected purely by chance. The probability of chance agreement for a single code is the product of the individual probabilities for each coder using the code. Kappa is then computed as $([a/a + d] -$ Chance agreement$)/(1 -$ Chance agreement$)$ where $a =$ number of agreements, and $d =$ number of disagreements. For the data in Table 3.4, Kappa equals 0.75 for the raw agreement of 0.8.

Because it takes chance agreement into account, Kappa is supposed to be a standardized measure. Harvey et al. [23] noted that data from extracting and coding items in text are "meaningless ... if coders are not reliable in their judgements," citing 0.7 as the critical value. Landis and Koch [24] provided a similar guide to acceptable Kappa values.* Unfortunately, Kappa has been extensively critiqued [25–27], and its use in this context is questionable. There are three main problems.

First, Kappa, because it is based on the index of concordance, does not provide a measure of *different types* of agreement or disagreement. For example, suppose the codes in Table 3.4 denote the following: Code A indicates a procedural problem, and Code E indicates a communication problem. Reporting that "agreement for the taxonomy was 0.8 with a corrected value of 0.75" obscures the simple fact that we had perfect agreement for procedural problem events but disagreed in half the events for which Alan or Beryl coded a communication problem.

Second, a problem with Kappa is that it is sensitive to prevalence (also called base rate [28]). This issue has been extensively discussed in the literature on psychiatric diagnosis and epidemiology [29]. In simple terms, the problem is that, even though it is the most commonly employed reliability statistic, Kappa does not just vary with reliability. This is evidently problematic and arises out of what Lee and Del Fabbro [30] called the *agreement bias*. To illustrate the problem, let us provide another small sample of ten events coded by our two coders, Alan and Beryl, using a five-category taxonomy that contains coding choices A, B, C, D, and E. The codes assigned this time are shown in Table 3.6.

---

* They argue that Kappa values of 0.81–1.00 indicate excellent or perfect agreement, 0.61–0.80 substantial agreement, 0.41–0.60 moderate reliability, 0.21–0.40 fair reliability, and 0.00–0.20 poor reliability.

TABLE 3.6
Codes Assigned by 2 Coders with 5
Choices to 10 Events

Event	Code Assigned		Agree (a)/Disagree (d)
	Alan	Beryl	
1	D	D	a
2	C	C	a
3	E	E	a
4	E	E	a
5	E	E	a
6	E	E	a
7	E	E	a
8	C	E	d
9	D	E	d
10	E	E	a

There are, as in Table 3.4, eight events that the coders agree on; that is, the raw agreement is 0.8 or 80%. However, in this case there is a decreased spread of events, with six agreements involving agreeing on Code E. Beryl used this code for eight of the ten events.

The Kappa coefficient for the data in Table 3.6 is computed as 0.58. When we compare this with the 0.75 we computed for Table 3.4, we see something strange has happened. In both cases, Alan and Beryl agreed on eight out of ten events, but there are two separate Kappa scores (in conventional terms, agreement in Table 3.4 is substantial, and in Table 3.6 it is moderate) [31]. This is because, for a fixed rate of agreement, Kappa coefficients will be higher when the proportion of agreements is spread across the choices available.

Because Kappa coefficients vary with coding distributions, the reliability of information on a human factors database will thus be contingent not only on the ability of investigators to agree, but also on what they actually agree on, which is clearly an unsatisfactory state of affairs [32]. It is important that safety managers can agree on common events as well as rare occurrences, and their ability to discriminate is our concern at this stage. Grove et al. [33] argued that Kappa is actually a series of reliabilities, one for each base rate. As a result, Kappas are seldom comparable across studies, procedures, or populations; that is, it is *not* actually a standardized measure.

Our fundamental problem with Kappa is that, like Lee and Del Fabbro [34], we do not accept the frequentist assumptions that underpin the statistic. Let us try to explain in simple terms what the problem is with using a frequentist measure in this context.

Suppose we have conducted a trial and obtained the data in Table 3.4. We concluded that raw agreement was 0.8 or 80%, and Kappa was 0.75 (calculated via the *Statistical Package for the Social Sciences [SPSS] for Windows*, version 11). This correction for chance agreement computes the chance of a given agreement as simply the product of the individual probabilities for each coder based on the number of choices available. The coders are also assumed to be entirely independent; that is, Alan's coding is not conditional on Beryl's or vice versa.

Now, suppose we conduct another trial and obtain the same results. Because the frequentist cannot assume any prior knowledge, that is, cannot look at a distribution of codes like those in Table 3.4 and say, "It looks to me that the chances of Beryl coding an E are about 4 in 10," a position of ignorance must be adopted (see above) by which a hypothetical random assignment of codes is used to generate chance probabilities. So, because there are five codes, the chance of Beryl coding an E at the start of the new trial, for the frequentist, is still 0.2, as it was for the first trial and as it would be for every subsequent trial. And, the two coders must still be assumed to be independent (despite the clear overlap in their codes in Table 3.4, in which 80% of the time they pick the same code). Kappa will be computed on the basis of these two assumptions.

The problem is, we think, that human beings do not conceive of chance agreement like this. We think the data in Table 3.4 would lead us to expect Beryl using Code E more than Alan in a subsequent trial. *We already know* Beryl uses Code E more than Alan, so why should we be forced to deny this? Also, we would hope that in a high-consequence industry there is more to the coding process than chance. It is hoped, if an event is about a scheduling problem, then there is more than a chance probability that Alan or Beryl (or hopefully both) will code it that way. How can two workmates coding events be entirely independent? They will have been trained together, they are both reading the same text or report, and they are *attempting* to code it in the same box. If they agree on a coding, then it is not a coincidence. It is what they are trying to do. If we assume *no* relationship between them, then it is as if I agree to meet you at lunch at 2:00 p.m., yet when we arrive together, we exclaim, "What are the chances of that!" because it is a big world, with lots of restaurants, and the probability of us being in the same one at the same time is objectively very low.

We prefer to use data like those in Table 3.4 to inform our beliefs about reliability (e.g., Alan agrees a lot with Beryl, Beryl likes Code E). We prefer not to use Kappa because, as a frequentist statistic, it uses concepts of chance agreement that we do not feel are particularly useful. Human beings do not use taxonomies as if throwing darts blindfolded or rolling dice; they try to assign codes using prior experience the best way they can, and (with a good taxonomy) more often than not they agree.

We think that the only way to define the chances of a coder using a particular code is to think about previous coding behavior, the conceptual aspects of the taxonomy, and so on. This Bayesian approach is codified in the BK statistic for coding agreement that we previously recommended [35,36]. But, there is a last proviso. Formal Bayesian statistics remain controversial, and perhaps more crucially, there is a pragmatic problem with them as their calculation involves working with standard statistical packages (e.g., *SPSS*). So, now we outline some easy ways of avoiding the traps inherent in frequentist statistics of reliability (e.g., Kappa coefficients) [37,38] by encouraging people to work at the level of simple conditional probabilities.

Simple conditional probabilities allow us to move beyond the overall (omnibus) agreement measures such as the index of concordance or Kappa coefficients and to start to explore more subtle biases in coding patterns. (We set out a similar argument in Chapter 4, encouraging a move away from the search for objective probabilities via standard probabilistic risk assessment tools or traditional tests of significance toward an exploratory or descriptive approach.)

3.5 SIMPLE CONDITIONAL PROBABILITY FOR TAXONOMIC CONSENSUS

Let us say that you are a safety manager who employs Alan and Beryl to code safety events into a database. You have worked out your taxonomy. It seems to have mutually exclusive codes and to cover all bases. You cannot see any bucket categories or any other obvious flaws, and you believe it ought to work because you have designed a taxonomy for your specific work situation because you know that off-the-shelf taxonomies will not work.

But, you still have the problem that you simply cannot get Alan and Beryl to agree on more than about 65% of codes no matter how hard you try. What should you do? We strongly believe that there has to be a preagreed standard, which we think should be set at 70%, that will leave everyone in no doubt regarding what needs to be achieved. So, a trial showing 65% agreement should be seen as a step on the road to a reliable taxonomy or a "close-but-no-cigar" moment.

To act, we believe there are more specific questions that must be asked beyond the question, what is overall agreement? An exploration of the *separate* conditional probabilities that underpin raw agreement should tell you what your course of action is.

Suppose we look at how Alan and Beryl have been using a single Code E over a small sample of ten events. In this case, for each event there are only two choices available to each coder: to code E or not to code E. The data observed are shown in Table 3.7. (To avoid confusion, we should remind the reader that although in a

TABLE 3.7
Codes Assigned by 2 Coders with 2 Choices to 10 Events

Event	Code Assigned		Agree (a)/disagree (d)
	Alan	Beryl	
1	E	E	a
2	E	E	a
3	E	E	a
4	E	E	a
5	E	Not E	d
6	E	Not E	d
7	Not E	E	d
8	Not E	Not E	a
9	Not E	Not E	a
10	Not E	Not E	a

reliability/consensus trial Alan and Beryl would be separated and not allowed to confer, they would, as in this example and the ones that follow in the chapter, be coding *exactly the same set* of events, texts, videotapes, etc., so that a direct comparison of codes applied to each event could take place.)

We can see from Table 3.7 that the raw agreement is right on the edge of respectability at 0.7 or 70%. But, simply reporting this may mask some important aspects.

First, the probability of Alan coding an event as an E [Pr(E^A)] is 0.6, and the probability of Beryl coding an event as an E [Pr(E^B)] is 0.5. The probability that they will *both* code an event as an E [Pr(E^A E^B)] is 0.4 (Cases 1, 2, 3, and 4 of ten trials).

From these probabilities, we can also calculate the *conditional* probability of Alan identifying an error given that Beryl does so. Here, we are only interested in events for which there is an E in the column labeled Beryl. This probability [Pr(E^A/E^B)] is given by Pr(E^A E^B)/Pr(E^B), which is 0.4/0.5 = 0.8 (i.e., there are five events that Beryl coded E, and Alan said E for four of them). Similar calculation will tell the chances of Alan *not* coding E given that Beryl *does not* do so (which becomes 0.6).*

The first thing to notice is that the chances of Alan agreeing with Beryl *depend* on what Beryl codes. If she codes E, Alan agrees with her 80% of the time, which is fine. But, if she were to code an event and *not* code it as E, then Alan's chance of agreeing with her is only 3/5 or 60%, which places us below the 70% threshold.

We do not think this is intuitively understood by safety managers and practitioners, who tend to report the overall reliability of a taxonomy and interpret this to be the chance of agreement on any given case or event (e.g., the usual interpretation of reliability data of 70% is that any coder has a 0.7 chance of agreeing with another on any given event). This is why moving beyond raw agreement is vital. What may also be counterintuitive is that the chances of Alan agreeing with Beryl's two choices that we have worked out (0.8 for E and 0.6 for Not E) are not the same as the chances of Beryl agreeing with Alan's codes. The chances of Beryl coding E given that A has done so [Pr(E^B/E^A)] comes out to 0.67, and the chances of Beryl not coding E given that Alan has not done so [Pr(Not E^B/Not E^A)] comes out to 0.75.**

So, we have a range of agreement from a 0.6 chance of Alan coding Not E if Beryl codes Not E to a high of 0.8 for Alan coding E if Beryl codes E. The chances

---

* The probability of Alan not coding an E [Pr(Not E^A)] is 0.4, and the probability of Beryl not coding an E [Pr(Not E^B)] is 0.5. The probability that neither will code E [Pr(Not E^A Not E^B)] = 0.3 (Cases 7, 9, and 10 of ten trials). Pr(Not E^A/Not E^B) is given by Pr(Not E^A Not E^B)/Pr(Not E^B), which is 0.3/0.5 = 0.6. Also, we can get Pr(E^A/Not E^B) (the chance of Alan coding E if Beryl *has not* done so) and Pr(Not E^A/E^B) (the chance of Alan not coding E if Beryl *has* done so), which come out as 0.4 and 0.2, respectively. (These tell us how *bad* Alan is at agreeing with Beryl's interpretations.)

** We get these inverse probabilities from Bayes theorem. We already know the probability of Alan coding one way or the other given Beryl's codes and their individual probabilities of coding each way. By Bayes, Pr(E^B/E^A) = [Pr(E^A/E^B)][Pr(E^B)]/Pr(E^A). This calculates to 0.67. Similarly, Pr(Not E^B/Not E^A) = [Pr(Not E^A/Not E^B)][Pr(Not E^B)]/Pr (Not E^A). This calculates as 0.75. Also, we can get Pr(E^B/Not E^A) (the chance of Beryl coding E if Alan has not done so) and Pr(Not E^B/E^A) (the chance of Beryl coding Not E if Alan codes E), which come out as 0.5 and 0.33, respectively. (These tell us how *bad* Beryl is at agreeing with Alan's interpretations.)

of Beryl agreeing with Alan are somewhere in the middle but are higher when Alan does not code E.

We feel it is vital that interpretations of reliability data take into account these different conditional probabilities. The error that we must avoid is generally called transposing the conditional; in this case, it is assuming the chances of Alan agreeing with Beryl to be the same as the chances of Beryl agreeing with Alan. Although $Pr(E^A/E^B) \neq Pr(E^B/E^A)$ may be axiomatic in probability theory (and therefore self-evident to mathematicians and statisticians), we are not aware of widespread understanding of this in reporting of reliability studies in psychology or in human factors and safety. This general feature of conditional probabilities will manifest itself in real-life coding work.

Suppose Alan and Beryl are coding a series of events. Now, every time Alan sees something in a report about a problem occurring after any information transfer, written message, verbal instruction, people having a chat or being spoken to or somebody being told something, something not being passed on, or anything similar, he tends to code the Communication Error code. In other words, he is a very *liberal* coder, at least as far as the communication code goes. Beryl, on the other hand, only uses this code when she actually sees the word *communication* used in formal terms (e.g., reference to a written document or at least a formal verbal instruction). In other words, she is a highly *conservative* coder as far as this code goes.

Alan will have more of these codes under his name overall, and if we know he uses the code, there will be a certain chance that Beryl will agree. But, if we know Beryl uses the code, we can make a much more confident prediction of agreement. Because Alan is the more liberal, any case Beryl says is a communication failure is very likely to be coded this way by Alan.

So, why is this practically important? Well, it tells us that disagreement appears to arise from a different conception of the code. We might want to say the following to Alan: "Look, you seem to be using this code very frequently. Could you possibly be a bit more skeptical about its use?" Or, we might say to Beryl: "Can you include less-formal examples under that heading?" Both approaches are likely to lead to more agreement. In fact, better still, we could get Alan and Beryl together (they may not be aware of their different interpretations) to talk about their use of the code and come to a consensus about how broad a series of events it should cover. Then, we could track their progress because it is hoped that their coding styles converged and reliability rose over the crucial 70%.

If we had only worked out an overall agreement level, then we could not have come to this conclusion about what to do. It may have been that they each used the code as strictly as the other, yet disagreed on specific events. The data in Table 3.7 are like this, with overall use of the code roughly equivalent for each coder. A better approach in this case would be to look at the taxonomy and the event reporting rather than the coders to see if there was information coming through that was difficult to classify using the codes we had designed.

To recap, raw agreement is a good indicator of how useful taxonomic work can be, and 70% is a workable threshold. Common corrections for chance are misconceived. But, raw agreement can mask clear directional bias. Sometimes, raw agreement may be acceptable overall but will mask specific cases for which agreement is poor by aggregating them with some for which it is very high. So, we encourage the working out of different conditional probabilities of agreement between coders.

Finally, we are conscious that these conditional probability equations can look a bit technical. The probabilities of Alan agreeing with Beryl when she codes an E (and when she does not) can, however, be described more simply as Alan's *sensitivity* and *specificity* to Beryl's codes. We are now going to turn briefly to the area of signal detection theory (SDT), in which these concepts were developed. Importantly, a technique has been developed in SDT that allows for sensitivity and specificity to be plotted in graphical form (this is a very simple procedure given access to a standard statistical package like *SPSS for Windows*). Thus, taxonomic agreement can be studied simply by looking at the shape of graphs, which we hope will appeal to people with an aversion to equations.

3.6 SIGNAL DETECTION THEORY AND RELIABILITY TESTING

SDT was developed during World War II for the analysis of the performance of radar operators. Radar operators had to decide whether a blip on the screen represented an enemy target, a friendly ship, or just noise. The theory developed because of concerns about the ability of radar receiver operators to make these important distinctions reliably.

The ability to make such yes/no decisions is called the receiver operating characteristic (ROC). This is a function of two aspects that can be identified from data observed. *Sensitivity* depends on correct positive identifications and decreases when a positive case is answered in the negative (called a *miss*). *Specificity* depends on correct rejection and thus decreases when a negative case is identified as a positive (called a *false positive*).

Examples of a sensitive ROC would include a diagnosis of depression among those with real depression or a valve problem found in events in which this really occurred. Specificity would be demonstrated by, for instance, *not* diagnosing schizophrenia in patients *without* the disease or *ruling out* a factor that in fact did *not* contribute to an accident.

Let us think of our two coders, Alan and Beryl, again and examine a series of codes that Alan assigns. Beryl's coding will be sensitive to Alan's if she assigns a code that Alan does. If she misses some of his codes (does not assign a code that Alan does assign), then sensitivity will decrease. Her specificity will depend on what she does when Alan does not use the code. If she uses the code (a false positive), then specificity will decrease, with her task in this case to match Alan's negative coding (correct rejection).

These twin measures are *exactly the same as* the conditional probabilities we worked through in Section 3.5. For example, the probability of Alan using Code E for a set of events given that Beryl also uses E [$\Pr(E^A/E^B)$] is Alan's sensitivity to Beryl's coding. His specificity depends on his ability to code Not E when Beryl codes Not E [i.e., $\Pr(\text{Not } E^A/\text{Not } E^B)$]. So, sensitivity is reduced with misses of a yes code, and specificity is reduced with false positives in relation to a no code.

The statistical computation of sensitivity and specificity for coding data presents no particular difficulty in itself because the contingency tables of an interrater reliability matrix and that of a classic signal detection matrix have the same mathematical properties.

There is one important aspect to remember. Sensitivity and specificity are always *directional*. For coding, we will always have sensitivity and specificity for each coder relative to the other one. In SDT, sensitivity and specificity scores are reported for a test against a source (or gold standard) that represents the true value. But, we have no such definitive account, so we want to work out Alan's agreement with Beryl's codes and Beryl's agreement with Alan's codes.

These agreements are based on the conditional probabilities we showed above and do not require a claim of objectivity for either coder. The comparison is still between decisions we assume to be subjective. Remember from the discussion on probability that that we would usually wish to calculate both directions of agreement in any case because they will usually not be the same.

A final data set may be useful to show how the plotting of sensitivity and specificity data can be illuminating. For simplicity, let us assume we have asked Alan and Beryl to code 100 safety events or cases and code them. We once more look at data pertaining to one code, Code E, as with the data in Table 3.7. This time, because we have 100 events, we have laid out in Table 3.8 the data as a contingency table that summarizes the four possible agreement cases (rather than presenting all pairs of codes assigned in list form as in Table 3.7).

We get the reliability of coding from Table 3.8 by adding the cases for which both say E plus the cases for which both say Not E and dividing by the total of 100 events. Reliability is $(12 + 32)/100 = 44\%$. (We would rightly be concerned with this.)

Looking now at sensitivity and specificity, we get Alan's sensitivity to Beryl's coding by dividing the 12 times Alan agreed with an E from Beryl by all 60 of Beryl's E codes = 0.2 or 20%. Because in this case Beryl is a far more liberal coder, Alan's sensitivity to her codes is low. But, Alan's specificity (how good he is at saying Not E when Beryl says Not E) is $32/40 = 0.8$ or 80%.

So, it would be wrong to look at the poor overall agreement of 44% and simply conclude Alan cannot agree with Beryl. Alan cannot agree with Beryl when she codes E, but when she codes Not E, he is rather good at doing so. He has a lot more

TABLE 3.8

Codes Assigned by 2 Coders with 2 Choices to 100 Events

		Beryl		Total
		E	Not E	
Alan	E	12	8	20
	Not E	48	32	80
Total		60	40	100

misses (48) than false positives (8). Alan is the conservative coder for this code; hence, when trying to agree with Beryl he is more specific than sensitive.

Beryl's ability to agree with Alan is also of interest. For the same overall agreement of 44%, she has a sensitivity to Alan's coding of 12 of 20 or 60% and a specificity of 32/80 or 40%. So, she is "the other way round." She does not miss so many of Alan's codes, but she quite often codes E when he does not (false positives), leading to lower specificity. Once more, we see a range in conditional agreement from 20% up to 80% for the overall score of 44%. Remember, liberal coders tend to be more sensitive and less specific when trying to agree with others.

The clear approach here, based on Alan's low sensitivity score and Beryl's low specificity score, would be to address the imbalance in prevalence of E codes in their coding (20 for Alan and 80 for Beryl).

The final thing we are going to do in this chapter is to show the plotting of the data from Table 3.8 in graphical form. We think this is a simple and elegant way of looking at reliability data for two coders and offers the possibility of recognizing reliability problems through a quick examination of the graphs.

3.6.1 ROC CURVES

ROC curves are simple plots of data of the type shown in Table 3.8. They involve plotting sensitivity against one specificity and have a basic purpose: to illustrate graphically how one source of data compares with another. Often, they are produced for a test to try to detect real cases of something, like a test for trisomy 21 (Down syndrome) trying to detect actual cases for which a fetus will or will not prove positive for the abnormality.

Here, we are interested in comparing one coder against another. So, our correctness is simply agreement, but there are no other differences in the calculation. The only difference is we will have two ROC curves, one for Alan trying to agree with Beryl (if you like, Alan is the test) and one the other way around (if you like, Alan is the source).

We know that Alan is the conservative coder and Beryl is the liberal one, and we want readers to be able to recognize this aspect simply by looking at the shape of the graphs. Let us plot the two ROC curves side by side now for comparison. Figure 3.2 is for Alan agreeing with Beryl, and Figure 3.3 is for Beryl agreeing with Alan.

It can quickly be seen from the shape of the graph that Alan has high specificity (graph stays close to the left-hand axis) and a lower sensitivity (graph falls down from top line) to Beryl's codes. It can also be seen from the shape of the graph that Beryl has lower specificity (graph moves away from left-hand axis) and higher sensitivity (graph goes close to top line) to Alan's codes.

The difference in shape between the two curves can be seen. Alan is the conservative coder; Beryl is the liberal coder. Or, Alan prefers to miss when getting coding agreement (errs on the side of caution) wrong, and Beryl prefers false positives (has a "scattergun" approach to using the code).

Looking at the shape of ROC curves, with a little practice and experience, tells us about the bias in coding of our two coders' agreement and the tradeoffs they

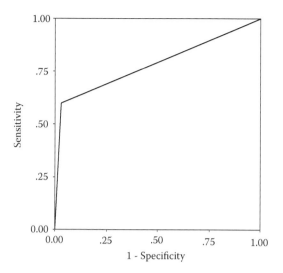

FIGURE 3.2 ROC curve for Alan agreeing with Beryl for the data in Table 3.8.

make so we can improve reliability. In this case, the reliability is 44%, so it will take a bit of improving. Remember, we are trying to train the coders to agree with one another in terms of assigning things (texts, events, audio data, videos, whatever) to categories that exist within a taxonomy. As we have seen that Alan is a much tighter coder (who needs more convincing to use the code), we may wish to tell him to "loosen up" a bit. And, as Beryl uses the code on whim without much persuasion,

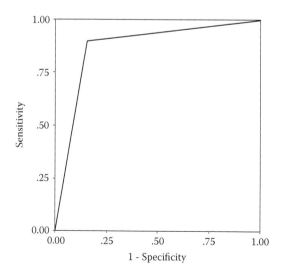

FIGURE 3.3 ROC curve for Beryl agreeing with Alan for the data in Table 3.8.

we may want her to "pull in the reins." Our next trial may then show how close we are to achieving adequate agreement via the 70% rule.

There are other practical benefits. Coders can be asked to recode events for an intrarater reliability trial. The basic measure of raw agreement will tell us whether there has been a shift in interpretation over time. Plotting the ROC curves here will show whether a coder is becoming *more* or *less* specific when using particular codes, that is, will pinpoint the source of the unreliability. In addition, if a new coder takes a job, we can use a trial to see how well he or she will fit into the established (hopefully reliable) coding of existing coders.

3.7 CONCLUSION

This chapter has been concerned with the reliability of taxonomies. Because of well-established problems with coefficients such as Cohen's Kappa, we have previously argued [39, 40] that the index of concordance [41] should be seen as the standard measure of agreement for safety data. This is simply the number of agreements divided by the number of coding attempts and may be multiplied by 100 to give a percentage agreement; it remains the industry standard for taxonomic work.

When we wish to explore beyond this overall level of agreement, we have argued that comparisons of coders' decisions can be made via an examination of simple conditional probabilities (probabilities of a coder assigning codes when other coders have or have not done so).

Finally, we have described how these probabilities can be understood (and displayed graphically) using principles from signal detection (namely, sensitivity and specificity) to give detailed and useful information about any biases that exist and that need to be addressed to achieve reliable coding. In this way, the different perspectives that coders bring to bear on things will not obscure the taxonomic data with which we are ultimately concerned.

We believe that what is known about a system (e.g., safety in a chemical plant) is bounded by consensus in people's interpretations of it (i.e., the taxonomic decisions they make about it). If a database or taxonomy (e.g., any human error coding scheme, human reliability assessment tool, root cause analysis system, minor event database, investigation methodology, etc.) is not reliable (i.e., users of the methodology are not able to demonstrate sensitivity and specificity when comparing coding decisions), then *it is not at all useful*. In fact we would go further and say any system of this sort for which reliability data are not readily available (i.e., in the public domain) exists to the detriment of safety, in that effort invested will, by definition, be more profitably employed elsewhere.

There is one final thing to say about the probabilities, sensitivities, and specificities and ROC curves with which we have been concerned in the latter half of this chapter. These are all intended to shed light on basic raw reliability data by splitting it into its component parts (e.g., who is using what code too much and why). But, the key factor remains whether we have crossed the finishing line we have set (our recommended 70% figure). The reliability trial itself produces the raw agreement, which is either above 70% or it is not. The other techniques (e.g., ROC curves, etc.) are data exploration techniques to pinpoint specific aspects of disagreement to work on so we can achieve reliability over the 70% threshold (or increase it to the highest level possible). This idea of the usefulness of exploring data sets is continued in the next chapter on database output.

REFERENCES

1. Groeneweg, J., *Controlling the Controllable: The Management of Safety,* 3rd ed., DSWO Press, Leiden, the Netherlands 1996, p. 134.
2. Stanton, N.A. and Stevenage, S.V., Learning to predict human error: Issues of acceptability, reliability and validity, *Ergonomics,* 41 (11), 1746, 1998.
3. Surowiecki, J., *The Wisdom of Crowds: Why the Many Are Smarter Than the Few and How Collective Wisdom Shapes Business, Economies, Societies, and Nations,* Doubleday, New York, 2004, p. 36.
4. Surowiecki, J., *The Wisdom of Crowds: Why the Many Are Smarter Than the Few and How Collective Wisdom Shapes Business, Economies, Societies, and Nations,* Doubleday, New York, 2004, p. 34.
5. Davies, J.B., Ross, A.J., Wallace, B., and Wright, L., *Safety Management: A Qualitative Systems Approach,* Taylor & Francis, London, 2003.
6. Ross, A.J., Wallace, B., and Davies, J.B., Measurement issues in taxonomic reliability, *Safety Science,* 42 (8), 771, 2004.
7. Wallace, B., Ross, A.J., and Davies, J.B., Applied hermeneutics and qualitative safety data: The CIRAS project, *Human Relations,* 56 (5), 587, 2003.
8. HMSO, *The Health and Safety at Work etc. Act,* HMSO, London, 1974.
9. Health and Safety Executive, *The Ionising Radiations Regulations,* HMSO, London, SI No. 1333, 1985.
10. Wallace, B., Ross, A.J., Davies, J.B., Wright, L., and White, M., The creation of a new minor event coding scheme, *Cognition, Technology, and Work,* 4 (1) 1, 2002.
11. Frohner, F.H., Evaluation of data with systematic errors, *Nuclear Science and Engineering,* 145, 3, 2003.
12. Lee, M.D. and Del Fabbro, P.H., A Bayesian coefficient of agreement for binary decisions, http://www.psychology.adelaide.edu.au/members/staff/michaellee/homepage/bayeskappa.pdf, 2002.
13. Davies, J.B., Ross, A.J., Wallace, B., and Wright, L., *Safety Management: A Qualitative Systems Approach,* Taylor & Francis, London, 2003.
14. Borg, W. and Gall, M., *Educational Research,* Longman, London, 1989.
15. Martin, P. and Bateson, P., *Measuring Behavior: An Introductory Guide,* Cambridge University Press, Cambridge, U.K., 1993, p. 120.
16. Davies, J.B., Ross, A.J., Wallace, B., and Wright, L., *Safety Management: A Qualitative Systems Approach,* Taylor & Francis, London, 2003, p. 128.
17. Borg, W. and Gall, M., *Educational Research,* Longman, London, 1989.
18. Davies, J.B., Ross, A.J., Wallace, B., and Wright, L., *Safety Management: A Qualitative Systems Approach,* Taylor & Francis, London, 2003, p. 126.
19. Ross, A.J., Wallace, B., and Davies, J.B., Measurement issues in taxonomic reliability, *Safety Science,* 42 (8), 771, 2004.
20. Cohen, J., A coefficient of agreement for nominal scales, *Educational and Psychological Measurement,* 20, 37, 1960.
21. Cohen, J., Weighted Kappa: Nominal scale agreement with provision for scaled disagreement or partial credit, *Psychological Bulletin,* 70, 213, 1968.

22. Fleiss, J.L., Measuring nominal scale agreement among many raters, *Psychological Bulletin,* 76, 378, 1971.

23. Harvey, J.H., Turnquist, D.C., and Agostinelli, G., Identifying attributions in oral and written material, in Antaki, C. (Ed.), *Analysing Everyday Explanation: A Casebook of Methods,* Sage, London, 1988, p. 32.

24. Landis, J.R. and Koch, G.G., The measurement of observer agreement for categorical data, *Biometrics,* 33, 159, 1977.

25. Maclure, M. and Willett, W.C., Misinterpretation and misuse of the kappa statistic, *American Journal of Epidemiology,* 126 (2), 161, 1987.

26. Feinstein, A.R. and Cicchetti, D.V., High agreement but low Kappa: The problems of two paradoxes, *Journal of Clinical Epidemiology,* 43 (6), 543, 1990.

27. Lee, M.D. and Del Fabbro, P.H., A Bayesian coefficient of agreement for binary decisions, http://www.psychology.adelaide.edu.au/members/staff/michaellee/homepage/bayeskappa.pdf, 2002.

28. Spitznagel, E.L. and Helzer, J.E., A proposed solution to the base rate problem in the Kappa statistic, *Archives of General Psychiatry,* 42 (7), 725, 1985.

29. Guggenmoos-Holzmann, I., How reliable are chance-corrected measures of agreement? *Statistics in Medicine,* 12 (23), 2191, 1993.

30. Lee, M.D. and Del Fabbro, P.H., A Bayesian coefficient of agreement for binary decisions, http://www.psychology.adelaide.edu.au/members/staff/michaellee/homepage/bayeskappa.pdf, 2002.

31. Landis, J.R. and Koch, G.G., The measurement of observer agreement for categorical data, *Biometrics,* 33, 159, 1977.

32. Davies, J.B., Ross, A.J., Wallace, B., and Wright, L., *Safety Management: A Qualitative Systems Approach,* Taylor & Francis, London, 2003, p. 132.

33. Grove, W.M., Andreasen, N.C., McDonald-Scott, P., Keller, M.B., and Shapiro, R.W., Reliability studies of psychiatric diagnosis: Theory and practice, *Archives of General Psychiatry,* 38(4), 408, 1981.

34. Lee, M.D. and Del Fabbro, P.H., A Bayesian coefficient of agreement for binary decisions, http://www.psychology.adelaide.edu.au/members/staff/michaellee/homepage/bayeskappa.pdf, 2002.

35. Lee, M.D. and Del Fabbro, P.H., A Bayesian coefficient of agreement for binary decisions, http://www.psychology.adelaide.edu.au/members/staff/michaellee/homepage/bayeskappa.pdf, 2002.

36. Davies, J.B., Ross, A.J., Wallace, B., and Wright, L., *Safety Management: A Qualitative Systems Approach,* Taylor & Francis, London, 2003.

37. Cohen, J., A coefficient of agreement for nominal scales, *Educational and Psychological Measurement,* 20, 37, 1960.

38. Scott, W.A., Reliability of content analysis: The case of nominal scale coding, *Public Opinion Quarterly,* 19 (3), 321, 1955.

39. Davies, J.B., Ross, A.J., Wallace, B., and Wright, L., *Safety Management: A Qualitative Systems Approach,* Taylor & Francis, London, 2003, p. 126.

40. Ross, A.J., Wallace, B., and Davies, J.B., Measurement issues in taxonomic reliability, *Safety Science,* 42 (8), 771, 2004.

41. Martin, P. and Bateson, P., *Measuring Behavior: An Introductory Guide,* Cambridge University Press, Cambridge, U.K., 1993, p. 120.

4 Taxonomic Output and Validity

In Chapter 2, we looked at the theory of building a taxonomy, how one might be created, how to assess whether a taxonomy was reliable, and why reliability mattered. In Chapter 3, we looked a bit more at the subject of reliability and discussed two views of probability: frequentist and Bayesian. Bayesian (or subjective) probability allows people to have prior ideas about what will happen (i.e., a subjective opinion on the probability of any given event; the opinion is then tested against later evidence). We argued that common reliability coefficients, based on frequentist ideas, are problematic, and that reporting simple conditional probabilities (sensitivities and specificities) is less contentious and describes better what we want to know. (We also think formal Bayesian statistics should be considered but recognize that learning how to use these is more time consuming.) In the simplest of terms, we think probability is a profoundly human concept used by and for human beings for pragmatic reasons.

If Chapter 3 was about input to databases, this chapter is about output. However, we continue the discussion of frequentist and Bayesian probability because, once more, *the most common techniques tend to be based on frequentist ideas*. We argue in this chapter that alternatives to frequentist data analysis are more philosophically coherent and easier to comprehend and provide more useful information. As in Chapter 3, we stop short of detailing fully blown Bayesian statistics and instead describe simple methodologies that can be employed. We should stress that this chapter does not attempt to be exhaustive: we do not attempt to discuss every available method of analyzing data. What we do is critique some common approaches and then outline a few techniques and methods that, we think, lead to more useful and meaningful output than that produced by the traditional methods.

4.1 TRADITIONAL ANALYSES AND POSSIBLE ALTERNATIVES

Now, as we saw in Chapter 1, in psychology and ergonomics, by analogy with Newton, human factors tend to be pictured as discrete (objective) entities (variables) with behavior that can be measured and predicted via context-free scientific laws. (We, of course, just see them as *taxonomic categories people apply*, reliably or otherwise, to bits of the system.) Further, it is assumed that an analytical/reductionist program can proceed by breaking up complex sociotechnical systems into these component parts, and that the system can then be understood in terms of these parts; Schön termed this view of science technical rationality [1].

What we demonstrate in this chapter is that frequentist ideas underpin the *statistical procedures* that are generally favored by those in the technical rationality tradition (e.g., probabilistic risk assessments, regression analyses, null hypothesis tests). Byrne [2] described such procedures as irredeemably linear. We discussed the limitations of the cause-effect metaphor in Chapter 1; at the moment, it is sufficient to say that these frequentist methods are used in the search for objective causality and they fit easily into the technical rational view of science.

Not satisfied with frequentist assumptions and following the leads of Byrne and John Tukey (especially Byrne's excellent text *Interpreting Quantitative Data* [1]), we stress in this chapter the importance of *descriptive* or *exploratory* statistics that avoid some of the pitfalls of the objectivist approach (which tends to focus on the testing of hypotheses and hence is sometimes also called the experimental approach). As Tukey put it:

> Once upon a time, statisticians only explored. Then they learned to confirm exactly — to confirm a few things exactly, each under very specific circumstances. As they emphasized exact confirmation ... , the connection of the most used techniques with *past insights* was weakened. Anything to which a confirmatory procedure was not explicitly attached was decried as "mere descriptive statistics," no matter how much we had learned from it. [3; italics added; note the Bayesian implication]

Later in the chapter (see Section 4.5), we recommend use of exploratory statistical techniques, such as cluster analysis (for generating a numerical taxonomy) and log linear analysis (for analyzing interactions in classificatory data). Both techniques are appropriate for analysis of reliable safety data of the type we talked about in Chapters 2 and 3, and we think they provide useful information. But, there is a wider rationale for recommending such analyses.

Remember, we are arguing against what we see as a *general approach to science*. We see a thread running through a technical rational philosophy, a search for true probabilities and a reliance on conventional statistical tests and objectivist methods of data analysis.

We are setting up here an *alternative general system* involving the prioritization of taxonomic hierarchies and recognition of the coherence of Bayesian views on probability and moving away from conventions of statistical testing toward a set of descriptive and exploratory alternatives.* We believe our approach will ultimately make everybody's life a lot easier, and that it is pragmatically and philosophically easier to justify than the traditional system.

To illustrate the problems with conventional approaches, we look at probabilistic risk assessment (PRA). A specific example discussed is an approach called a failure modes and effects analysis (FMEA). The problems inherent in such approaches, we argue, are indicative of the problems of technical rationality and frequentist (objectivist) probability as a whole. Then, we look at the null hypothesis test, which is probably the most commonly used and reported device in current science and is

---

* This should be a reminder of the Goethian approach discussed briefly in Chapter 1. Goethe's approach (and the exploratory approach here) looks *at* the data rather than *through* them and explores the data rather than using them to test hypotheses.

prevalent in all leading human factors journals. We argue, as others have in the past, that use of this test obscures the issues of real scientific and practical importance. Finally, we discuss alternative methods of analysis that exemplify the taxonomic (Bayesian in the broadest sense) and exploratory approach.

4.2 PROBABILISTIC RISK ASSESSMENT

To begin, let us look at PRA (alternatively termed probabilistic safety assessment, PSA). (Under the umbrella of PRA, we can include work on Human Error Identification [HEI], FMEA, human error probability, and the like.) Let us start with what is generally considered to be the first major PRA.

The Atomic Energy Commission commissioned a report in 1972 that has come to be accepted as a landmark in risk analysis [4]. The seminal nature of the report is perhaps indicated by the fact that there are various shorthand or informal ways to describe it, such as simply the Reactor Safety Study (RSS) or the Rasmussen report (after its author Norman Rasmussen, who was at the time an engineer at the Massachusetts Institute of Technology in Cambridge, MA). The report is generally thought of as the first modern risk analysis [5]. The report was controversial for a number of reasons [6], but the main scientific controversy was over the methodology used to calculate risks, specifically the use of expert judgment or subjective probabilities.* Miller pointed out why the RSS generated such controversy: "It violates the preference for the objective over the subjective, for the quantitative over the qualitative, and for *Frequentist* over *subjectivist* approaches to probability" [7; italics added).

Because of this controversy, the House Committee on Interior and Insular Affairs requested a formal external review, often called the Lewis report after its chairman, the physicist H. W. Lewis [8]. The Lewis report concluded that the RSS

> is defective in many important ways. Many of the calculations are deficient when subjected to careful and probing analysis, with the result that the accuracy of many of the *absolute probabilities* calculated therein is not as good as claimed. ... The *absolute values* of the risks presented by the Report should not be used uncritically either in the regulatory process or for public policy purposes. [9; italics added; note the frequentist paradigm]

Perhaps the question that will be troubling readers is, given such damning comment on the Rasmussen RSS report, why it is generally accepted as a landmark report in risk analysis? Perhaps the best answer is that it was praised not so much for what it *achieved* as for what it was *trying* to achieve. The Lewis report in particular praised the RSS for (apparently) advancing attempts to estimate risk and for providing a logical framework for doing so. The terms *absolute probabilities* and *absolute values* are not what was controversial (unfortunately); rather, the fact that

---

* In the previous chapters, we touched on the notion of the privileged expert. We asserted that useful estimates are usually socially agreed rather than determined by prioritizing of an individual expert account. Nevertheless, readers will no doubt guess that what we object to in PRA is not its use of subjective judgement per se.

some subjectivity in the process was identified (as if this could ever somehow not be the case) was controversial.

This RSS framework consisted of applying methods known as fault-tree analysis and event-tree analysis. (Event-tree analysis traces consequences of an event, and fault-tree analysis traces causes of a failure; the taxonomic nature of constructs like event and failure is evident.) Both use graphical tree forms of representation, with binary branching (the valve fails or does not fail; the core melts or does not melt; the containment holds or does not hold) at each branch point, where the probabilities of each alternative are determined. The probability of operator error or maintenance error can also be included, as well as the probabilities of casualties and damages, for example, if radioactivity is released. Thus, the probability of a particular chain of events can be calculated via the probabilities of all contributing alternatives.

The key question is, what is the source of the numbers (the probabilities)? For the most part, they come from what is coyly referred to as expert judgment. Or, as von Hippel later put it in less diplomatic language: "When crucial numbers were not available, they were simply guessed" [10]. In an article that set out the history of PRA since the RSS in some detail, Miller [11] noted that "The use and justification of expert opinion in risk assessment indeed became a central point of contention in the subsequent development of PRA, one that still has not been fully resolved."

The report made the use of subjective probability explicit. Lewis was cautious about the results that can be produced in this way:

> It is true that a subjective probability is just someone's opinion. But ... some people's opinions can be very accurate, even in a quantitative sense. ... For many of the steps in which a subjective probability was used it was the output of experienced engineering judgment on the part of people familiar with events of that type. This, of course, does not guarantee the accuracy of the probabilities so generated, but if properly chosen makes them the best available. [12]

The term *accuracy* gives the game away. The crux of the matter is that the entire discussion takes place within the context of an objectivist (frequentist) view of probability. The context is "true probabilities exist, therefore there is some concern about subjective estimates, but sometimes these are seen as the best we can do."

Although we feel that the idea of an objective (context- and interpretation-free) safety science (or any science) is a little ridiculous, many people have argued (and still argue) that this should be our ultimate aim. Thus, many subsequent authors spoke out to *criticize* the use of subjectivity in PRA [13–17]. Redmill described how, when a method involves "probabilistic equations for determining the likelihood of ... hazardous events" and when data for input into the equations are sparse, people may admit that they "just employ an expert to provide an opinion." He noted that "A computing acronym, GIGO (garbage in, garbage out), is also appropriate to mathematical risk models, but the results of risk analyses are often taken to be accurate and the assumptions and inaccuracies in their derivation unrecorded and forgotten" [18].

The Human Factors Reliability Benchmark Exercise (HF-RBE) of 1989 [19] was carried out on behalf of the European Commission to review the state of the discipline of human reliability assessment, which is a major PRA strand for safety managers and, obviously, the type of PRA most interesting to psychologists. Major techniques such as Success Likelihood Index Method (SLIM) [20] and Technique for Human Error Rate Prediction (THERP) [21] were evaluated. The conclusion was that the final probabilities derived depended on assumptions made by analysts, and that, as Hollnagel put it, "None of the methods used in this exercise were robust enough to compensate for the differences between individual analysts and analysis teams" [22].*

Now, the crucial issue is that this basic strategy laid out in the RSS has been at the core of safety management ever since and will be familiar to anyone who has attended a conference or seminar in the field. When the official investigation of the Three Mile Island nuclear accident subsequently endorsed the use and development of PRA techniques, the scene was set for a major explosion (no pun intended) in work of this nature.

The fundamental goal of PRA is to determine an objective (frequentist) probability. But, crucially, the numbers input are subjective estimates or are derived from classifications that are subjective. Either probabilities are simply put forward by judges, or probabilities are calculated from ratios of events that are defined by (framed by) subjective classification (see Chapter 3 on how we can never have an objective reference class to derive probability).

The PRA approach is flawed because the Bayesian aspects to the process cannot be reconciled with the underlying philosophy. Hence, there is criticism of its subjectivity from those who assume some other form of probability (an objective one) is possible to obtain. When the RSS was praised, it was because it was seen as the beginning of an effort to provide absolute probabilities. We hope this is clear; the RSS was criticized for including subjective judgments but praised for providing a means for logical progression by which objective probabilities could ultimately be achieved.

So, the (still hugely prevalent) PRA approach was born with a defect; despite the centrality of subjective judgment in the seminal study, people who picked up the mantle held fast to the goal of giving objective probabilities for events, and indeed objective (frequentist) probabilities have been sought like some sort of Holy Grail in safety science ever since.

We think the Lewis investigation was a product of its time. Although the reporters were able to recognize the subjectivity in the RSS (and by extension in PRA), they were writing at a time when objectivity about risks was still thought to be the aim of any serious risk science, and they could not "grasp the nettle" of subjectivity in the way we attempt here.

To be fair to Lewis, however, many contemporary commentaries still start from a position by which an objective probability of an event is deemed (a) possible and (b) desirable. Subjectivity and individual differences are still viewed by some as problematic. In some circles, it simply does not do to suggest a science that recognizes subjective probability.

---

* It is also interesting to note that the HF-RBE criticized human reliability analysis in general because of the assumption that human performance can be studied in much the same way as the study of technical aspects of a system: "Man is not a machine and complex interaction cannot be easily decomposed and modeled deterministically into a structure of elementary actions without losing subtle feedback, feedforward and other dependency mechanisms" [19, p. 118]. This aspect is addressed in Chapter 6.

In summary, the Rasmussen study, from our current perspective, involved Bayesian probabilities (estimates or degrees of belief about possible events). But, because of the ubiquitous nature of the rhetoric of technical rationality, subsequent use of PRA has been dogged with controversy over this aspect. Because the RSS was so openly criticized for its use of subjective probabilities, those involved in PRA/PSA tend to mask or play down the subjectivity in the methods as if this is something that is shameful [23].

We believe that PRA ultimately will be doomed to failure if its goal is an objective (or ultimate) probability because probability in science is better conceived as the degree of belief that something will happen. PRA generates estimates (or guesstimates) of what may or may not happen in a given system, and these rely on subjective input (either in framing the calculations by classifying events or cases or in providing probabilities directly) from specific people in a specific context. The search for a single true probability via a PRA we feel is like the search for a square circle or a perpetual motion machine; that is, the quest itself is fundamentally incoherent and misconceived.

4.2.1 Failure Modes and Effects Analysis

A good example of this confusion over the nature of probability in PRA is the FMEA technique. This traditionally involves inputting a subjective severity rating (numerical ranking of the effect of a particular failure mode, ranging from minor to catastrophic); a subjective probability factor (numerical weighting of causes identified for failure modes ranging from highly unlikely to inevitable); and a subjective likelihood of detection (estimate of how likely controls or barriers are to detect the cause of the failure mode and prevent an incident).

Once these estimates are made, a final risk priority number is generated objectively via a simple algorithm. But it is rare to see the subjectivity of the process explicitly recognized. There is nothing absolute about the ratings, weightings, and likelihoods involved. The best we could say might be that they were reliable across different trials (if this proved to be the case, see Chapter 3), but they are subjective probabilities nevertheless.

We want to make it clear, therefore, that we agree with Redmill, who noted the following: "Many models for human reliability assessment (HRA) are probabilistic. However, although probabilities are derived, the approach taken in most cases is based principally on human judgment. The results are at best reasonable approximations, and at worst wild guesses, but *always they include considerable subjectivity*" [24; italics added].

Now, we argued in Chapter 3 that using subjective probabilities revised in light of incoming evidence seems to be a very sensible way of approaching probability and science. The objectivist position that defined science when the RSS was carried out meant its use of expert judgments was seen as a flaw. But, notwithstanding what we said in Chapter 2 about experts (we return to discuss this in Chapter 8), we would certainly argue that judgment based on experience is precisely what science is about. The ideas that there can be knowledge *without* subjective judgment, or that we should approach scientific issues by simply pretending we have no prior assumptions or experiences that will be brought to bear are notions that we have been at pains to refute throughout this book.

We hope readers will see that the history of PRA is defined by the same historical tension between objective and subjective probability (and therefore science) that we discussed in Chapter 3.

4.2.2 Final Thoughts on PRA

To recap, PRA has been hampered by the fact that subjective estimates of probability are always required at some stage in the process, yet this requirement is normally treated like some form of dirty secret rather than made explicit and incorporated within a Bayesian framework.

Designers of PRA tools have, in essence, by use of databases of previous event data and the like, attempted to iron out the subjectivity in the process. This has been to little avail: by 1990, Apostolakis was still able to write that "no PSA [PRA] has been performed to date that does not use subjectivist methods (although very few analysts state explicitly that they are using Bayesian methods)" [25].

We should point out that those who argue they do not employ expert judgment but obtain objective data instead fall into a similar trap. Alternatives to subjective judgment are sometimes thought to be available because of the same misconception about the existence of a true probability. Hollnagel [26] offered the choice of empirical data or data from simulators. The empirical data are divided further into data from incident reporting systems or human error rates. These human error rates are "ideally determined from error frequencies as the ratio between the number of errors of a given type and the number of opportunities for that particular error." It is hoped that readers will remember the problem of reference class we mentioned in Chapter 3. Defining a particular type of error and the associated opportunities for that error is always a matter of *choice*. This is not an alternative to subjective judgment; it merely locates the subjectivity one step back in the process.

There is no objective probability based on a ratio of one event to another because we first have to obtain social agreement on what constitutes each event (this is the same problem as we faced in the bus crash example in Chapter 3). As Redmill [27] put it: However accurate the technical aspects of a risk analysis may be, the results are predisposed, or distorted, by the definition of the terms of reference, the exclusion of certain types of evidence, the definition of the study and system boundaries, and, in general, by the strategic planning of the analysis itself."

Desrosieres [28] called subjective probability epistemic. David Byrne [29] was clear that the epistemic is concerned with probability as the basis for action, while "the Frequentist is Platonist and contemplative," and his or her "knowledge of the world ... is valued for its own sake and according to its accuracy." (The links between this view and technical rationality, with its Platonic roots, could not be made clearer.)

This to us is the crux of the matter. We feel safety engineers do not want to know what causes events or whether slips, mistakes, or lapses abound in their workplace *for the sake of it*. Again, this is safety science aping the methods of pure science when those of applied science would provide far more useful data. In fact,

Byrne [30] went further and said the difference is between pure scientists and engineers. So, there we have it, we are taxonomic engineers, interested not in knowledge for its own sake but for discrimination, prediction, and action.

Unfortunately, the quest for a more and more objective PRA has, in addition to futility, obscured the real goal of testing models that we can use to make decisions. We return to this at the end of this chapter. First, we must discuss another traditional (frequentist, technically rational) methodology we regard as fatally flawed: the use of null hypothesis or significance tests.

4.3 PROBLEMS WITH THE NULL HYPOTHESIS TEST

Byrne, whose text we continue to draw on heavily here, compared and contrasted our two competing views of probability using the specific example of the null hypothesis, which is familiar to most students of science. Its widespread use is perhaps the most widely felt implication of the dominance of objectivist ideas. As Edwards et al. put it when criticizing its use: "No aspect of classical statistics has been so popular with psychologists and other scientists as hypothesis testing" [31].

Edwards et al. were writing in 1963, and we should therefore start with making something clear: *criticism of the null hypothesis test is not new.* This may have a vaguely familiar feel to psychologists, as almost all articles in which the test is discussed in critical terms begin by admitting that such criticism has been around for years. For example, much of what is said here can also be found in a famous article by Jacob Cohen [32]. We are indebted to Cohen and have quoted from his article a number of times. But, Cohen in turn quoted from, among many others, David Bakan [33], Paul Meehl [34], and Bill Rozeboom [35], and indeed started by saying, "I make no pretense of the originality of my remarks in this article." At the risk of overstating the case, we could mention that Bakan himself started in 1967 with, "Little of what is contained in this paper is not already available in the literature."

The reason we continue to bang this particular drum is simply that, despite decades of cogent argument against its use, null hypothesis testing is still hugely prevalent, something that should concern all of us. This section may appear to be a bit of a one-sided rant, but as Rozenboom put it, "One can hardly avoid polemics when butchering sacred cows."

4.3.1 WHAT THE TEST DOES IN THEORY

The null hypothesis test (also generally known as the significance test) has a simple aim: to gather information about a system that is not possible to observe in its entirety (normally because it is too big). This is done by extrapolating (the technical term is *inferring*) from something that it *is* feasible to observe: the sample (S). The job of the test is to help the scientist make inferences from the observed sample to the unobserved whole system (called the population [P]).

The *null hypothesis* is a specific proposition about the value of something in the sample (the something is formally called a *parameter*). First, we observe some data in the sample. The test involves calculating the probability (this probability value is usually referred to as simply the p value) that the data would recur in the population (strictly speaking, from another similar sample drawn from the same population) *if* the specific proposition we have made (the null hypothesis) were true. If the *p* value is below a certain level, we reject the null hypothesis and say we have a significant result. If the *p* value is above a certain level, we fail to reject the null hypothesis and have a nonsignificant result.

Now, let us try and be very clear about this. The *p* value relates to the probability of the data observed recurring in another trial drawn from the population if the specific case setup (the null hypothesis, or *Ho*) were true.

The point of the test is also very clear. It is to reject the null hypothesis or to fail to reject it. This strict dichotomy requires an arbitrary level (sometimes called the *α level*) to be set. If below the level, then we reject the hypothesis; above the level, we fail to reject. The level is often $p < 0.05$ (1-in-20 chance that the observed results would reoccur in repeated samples if the null hypothesis were true). This level, the critical threshold, actually dictates the chance of rejecting the null hypothesis *Ho* when it is in fact true (called a Type I error, finding a significant result by mistake).

For example, suppose we are interested in the relative performance of our male and female operators. We have a large organization, so we cannot test everyone. We pick a sample of men and women, match them for age and experience, and grade them out of 100 on a performance task.

Now, we are actually interested in what differences there are in performance between men and women. But, we approach this rather obliquely by imagining for a moment the null hypothesis, which is a specific point estimate about performance in our population. The null hypothesis in this case would almost always be that the mean performance of men and women is identical.* The test generates our *p* value, which tells us how likely it would be that we would obtain similar data from another similar test sample from our population given that the null hypothesis is true (if the women really are the same as the men).

Suppose $p < 0.05$, the critical value, and we reject the null hypothesis. The inference we are allowed to make about the population is simply that we would be unlikely to get another sample of data like these *if it were true* that women were exactly the same as men in this respect.

This is the classical (frequentist, technically rational) approach to data analysis. Let us now outline the problems with use and interpretation of the test that led Bill Rozeboom to say of null hypothesis testing as far back as 1960 that "the statistical folkways of a more primitive past continue to dominate the local scene" [36].

4.3.2 Confusion about Probabilistic Reasoning

A central concern about the null hypothesis test is that, as Cohen [37] showed, there is confusion over the probabilities involved, and this follows from its objectivist or frequentist assumptions. In our performance example, for instance, the frequentist has to believe that there either is or is not a performance difference between men

---

* In Section 4.3.3, we point out that the null hypothesis can actually refer to any point estimate, for example, that women are five performance points better than men, but that, in practice, the hypothesis of no difference is almost universally employed.

and women (remember this is, for him or her, an objective fact about the world). The real (absolute) probabilities of an identical mean performance on the task for men and women are 1 (identical) or 0 (not identical).

The problem is that this is not generally understood, so people tend to assume that the null hypothesis test gives us the probability that male and female scores are identical (further, it is assumed that the reciprocal probability equates to the probability that they are different). So, if the *p* value is less than 0.05, then it is assumed that there is a less than 5% chance that the scores are the same and then assumed that there is a 0.95 probability that the scores differ.

It is time for the slightly tricky, yet crucial, portion. The problem is that, as we saw in Chapter 3, it is illegitimate to transpose the conditional in probability. We noted that $\Pr(A/B) \neq \Pr(B/A)$: the probability of A given B is not the same as the probability of B given A.

The interpretation of the test we have just outlined is flawed because it comes from transposing conditional probability. The test asks us, as we have outlined, *Given that* Ho *is true,* what is the probability of these data? But this is not what we tend to be interested in as scientists. What we *really* want to know is, *given these data,* what is the probability that *Ho* (the null hypothesis) is true? [38]. So, people tend to assume (this is an act of wishful thinking rather than a complete accident) that the *p* value (which is the first case) refers to the probability of the null hypothesis being true (the second case). As we showed in Chapter 3, this is a total misconception.

Thinking back to our example, what we want to know is if the performance of men and women differs. What we actually found out was that it was not likely that the data from our sample would arise in another sample from the same population if it was true that performances were identical. Here, people become confused and think the test has shown that it is unlikely that the performance of men and women is the same. If we do this, we have confused the probability of the data given *Ho* is true [$\Pr(D/Ho)$] with the probability of the *Ho* being true given the data [$\Pr(Ho/D)$]. Remember, this is like concluding that the probability of someone who is French being 5 feet 11 inches tall is the same as the probability of someone who is 5 feet 11 inches tall being French.

The test has no bearing on how likely it is that the performance of the men and women is the same or different. Remember the frequentist's problem: in the frequentist's eyes, this proposition is either true or false; there can be no subjective estimate about its likelihood.

We, the testers, set up at the beginning the *Ho* that men and women were the same. The fundamental paradox is this: we cannot use a *Ho* (significance) test to find out how likely our null hypothesis is. As Cohen thus concluded of null hypothesis significance testing (NHST): "What's wrong with NHST? Well, among many other things, it does not tell us what we want to know, and we so much want to know what we want to know that, out of desperation, we nevertheless believe that it does!" [39].

In Chapter 3, we said the frequentist model allows for no probability to be set for something that is supposed to be objectively either true or false, that is, for no subjective opinion on whether the null hypothesis is true. The likelihood of the null hypothesis at the end of the trial [$\Pr(Ho/D)$] is actually a posterior probability and can only be derived from a Bayesian approach that takes into account how likely it was *before* the trial. If we knew this prior, we could estimate the posterior via Bayes theorem. Because this Bayesian probability is often claimed by authors to have been determined by the frequentist test, Gigerenzer called this type of confusion the "Bayesian Id's wishful thinking" [40].

The poor frequentist, constrained by the strict interpretation of his or her results, which only show a probability of data observed given a particular hypothetical case for "a theoretically infinite number of repeated exercises in which a sample of the size of our sample was drawn from a population" [41], is seduced by the common sense notion of Bayesian probability and comes up with an unjustified degree-of-belief type probability, thus becoming hopelessly lost in a pseudo-Bayesian fog.

Unless we embrace Bayesian thinking, we cannot say that the *Ho* is *probably not* true (subjective estimate) just because we have found these specific data are improbable (under certain conditions) if it *was* true. Whew!

4.3.3 INFERENCES FROM REJECTION OF THE *Ho*

Another common assumption that adds to the confusion is the strange idea that there are only two possible theories about any given phenomenon in the world, the *Ho* and your own *particular* view, and that if you reject *Ho*, then you can somehow prove the other.

For example, suppose a manager believes she has a training program that will improve reaction times because it is better than the existing program. She sets up a trial with new recruits. Two groups of people are involved, with Group A given the existing program and Group B given the new program. A *Ho* is set up that there will be no mean difference in reaction times for the two groups. The *Ho* is rejected at $p < 0.001$. The manager now goes to a safety meeting, reports a highly significant result, and argues that "it has been clearly shown that the new program is better that the existing one." Note how easy this conjuring trick seems.

Of course, all you have done is *reject* a thesis that (to reiterate) was most probably false from the outset. It is invalid to draw any further inferences from this. You cannot assume that because you have falsified Thesis 1 that Thesis 2 (or any thesis) is therefore *true*. There may be any number of theories that also explain the data. All you can do is go on to try to falsify them as well. NHST cannot be used to make positive statements about truth; all it can do is reject theories, *not prove them*.

You can only assume that Theory 2 is correct if you have falsified Theory 1 if there are logically only two possible theories. But, usually, and certainly in the example above, there might be all sorts of reasons why the reaction times were different in this particular case; it is invalid to infer that an idea you have about what caused the results is proved.

You will have to take our word for it, but this assumption (i.e., that by rejecting the null hypothesis you have therefore proved something or other) is far more common than you might think, prompting many authors to stress that statistical significance and scientific significance (i.e., theoretical importance) are independent [42]. For example, in reference to clinical trials, Gardner and Altman [43] argued that statistical significance should *not* be confused with medical *importance* or biological *relevance*. We agree with Zeisel, but feel he was perhaps rather cruel in saying that "the researchers who follow the statistical way of life often distinguish themselves by a certain aridity of theoretical insights" [44].

4.3.4 THE ABSURDITY OF THE NIL HYPOTHESIS

A further issue is that the *Ho* is almost always set up as a hypothesis of no difference. In fact, the *Ho* can formally be any specific case (e.g., that there is a five-point difference in some parameter between two groups or whatever). However, the test of the no difference case is so universal that Cohen termed *Ho* the nil hypothesis [45].

Cohen stated: "Things get downright ridiculous when *Ho* is to the effect that the effect size (ES) is 0, that the population mean difference is 0, that the correlation is 0 ... that the raters' reliability is 0" [46]. Now, correlations of exactly zero will not be something many readers can remember in *any* operational setting. As Edwards et al. famously put it: "In typical applications ... the null hypothesis ... *is known by all concerned to be false from the outset*" [47; italics added].

We believe such criticisms have a ring of common sense to them. If we really had an idea that there was *no* difference in something, would we be testing it? Can we imagine a single situation in our organization in which a variable (e.g., levels of fatigue, types of error, features of communication, etc.) would not vary *at all* across staff groups, shifts, work details, and so on? Or would we test, for example, two training programs that we believed led to *exactly the same* outcomes? Remember, in the example, people investigating differences between two training programs would almost universally set up the *Ho* that the results from each were *identical.*

It has often been argued that this is absurd. Bakan [48] was clear that "a glance at any set of statistics on total populations will quickly confirm the rarity of the null hypothesis in nature." Nunnally had another way of putting this: "If the null hypothesis is not rejected, it is usually because the N (*size of the sample*) is too small" and concluded that, "If rejection of the null hypothesis were the real intention in psychological experiments, there would usually be no need to gather data" [49].

Historically, *Ho* tests have their roots in Popperian philosophy. Karl Popper proposed the falsification thesis by which science could proceed by attempts to falsify theories. These could not be proved but could be strengthened by repeated failures to falsify them [50]. This dichotomy of falsification versus failure to falsify is similar to the rejection versus failure to reject distinction in *Ho* testing. You cannot prove a *Ho*, only fail to reject it.

But there is a fundamental problem with Popper's theory: science does not actually proceed via falsification per se. If this was true, it would by definition be advanced when *ridiculous* theories were falsified. So, scientific progress would be made if a scientist began his or her day by hypothesizing that the moon is made of green cheese and then attempting to falsify this statement. Because we intuitively understand that science would never progress under these circumstances, most philosophers (and scientists) adopt a modified version of Popper's theory: science proceeds by falsifying *plausible* theories.

If a null hypothesis is plausible (i.e., predicted by a respected theory), then we have a different case entirely. Indeed, Cohen, a fierce critic of NHST, argued that it *is* valid to set up a null hypothesis predicted by a (serious, plausible) theory because rejection of the *Ho* in this case would challenge the theory proposed. Meehl called this the strong form of Ho testing; this variety is theoretically beyond reproach [51, 52]. But, such NHST is very rare. Instead, as we have seen, people attempt to prove their own alternative theories by falsifying a hypothesis that is predictably false (usually the nil hypothesis). This is the "falsifying the idea that the moon is made of green cheese and therefore assuming you've proved it to be made of papier mache" version of science. We hope this example shows the absurdity of approaching this task in this way.

4.3.5 STATISTICAL POWER, EFFECT SIZES, AND CONFIDENCE LIMITS

Now, Nunally's reference to "the sample size being too small" indicates another Achilles' heel of the null hypothesis: if you do not succeed in rejecting the null hypothesis, then it is usually possible simply to gather more subjects until the *p* value falls below the critical level. The more cases we have (people, events, etc.), the greater the power of the test to pick up small differences in the population (called effect sizes).

Informally, we are well aware that people (unfortunately) know that if their study is not showing the "correct" result, they simply have to print off a few dozen more questionnaires and get them returned; with this larger sample size, they will soon get a "result" (assuming the variance stays the same).

This notion of sample size affecting the final decision led Thompson, tongue in cheek, to say: "Statistical significance testing can involve a tautological logic in which tired researchers, having collected data on hundreds of subjects, then conduct a statistical test to evaluate whether there were a lot of subjects" [53].

Criticisms of the null hypothesis test on this basis are not at all new: as Bakan said rather succinctly in 1967, the fact that "the probability of getting significance is also contingent upon the number of observations has been handled largely by ignoring it" [54], or in Cohen's analogy, "this naked emperor has been shamelessly running around for a long time" [55].

Now of course, this is a well-known problem, and responses have been created. The most common solution to the problem is to encourage researchers to report *effect sizes* [56].* The logic here is simple. Because *Ho* is likely to be rejected given enough cases, differences will be significant. But, this does not mean they are important. A *tiny* difference (effect size) can be significant if we have a high enough number of subjects.

The effect size is therefore an attempt to show whether the result of an experiment or study is *scientifically* significant rather than just *statistically* so and is certainly a step in the right direction, moving us beyond the strict reject/fail-to-reject decision that is the purpose of the *Ho* test toward something more meaningful. Effect sizes are typically reported with upper and lower confidence limits. The mean difference in a sample is a point estimate (i.e., a single number).

---

* There are different measurements for effect size, but the principle is the same: to move beyond statistical significance (rejection of a null hypothesis) and report on scientific significance (inferences about population parameters).

Although we agree that effect sizes are an improvement on simply reporting whether $p < 0.05$ (because they are more *descriptive*), there are still crucial problems. First and most mundane, effect size reporting remains very much the exception to the rule. For example, studies after the American Psychological Association recommended effect size reporting [57] have tended to conclude that this has had little impact [58, 59]. (One such study by McLean and Ernest also interestingly concluded that science is subjective, and this form of statistical test cannot make it any less so [60].)

Second, the frequentist logic we talked about in Chapter 3 still applies to effect sizes. The frequentist cannot say how *certain* he or she is about whether the true value in the population is within the confidence range (as it either is or is not). Only Bayesians can give a subjective estimate stating how *confident* they are that an effect size is within a certain range (called a Bayesian credible interval).

Finally, it has been pointed out that reported effect sizes tend to be inflated. There are two reasons for this. First, the *publication bias* means statistically significant results (which by definition have bigger effect sizes, all other things being equal) tend to be reported and nonsignificant ones ignored or repressed [61]. In other words, if your experiment gets a (significant) result you are more likely to publish than if it does not. So, the received wisdom on a particular scientific issue tends to be based on a partial picture distorted in favor of bigger effect sizes.

Perhaps more interestingly, it has been shown that, irrespective of what is subsequently reported, the effect size needed to obtain a (statistically) significant result in a sample tends to be bigger than the effect that may exist in the population from which the sample is drawn. For example, Schmidt [62] outlined a case for which there is an effect size of 0.5 in a (whole) population. He showed that if you have a random sample of 15 subjects from this population, you need an effect size of around 0.65 to get a significant result.

Obviously, a sample of the whole population (if that is not a contradiction in terms) would show an effect of 0.5. But, as a sample has (by definition) less discrimination (power) than the larger population, a bigger effect is needed before this shows up as significant. Sampling theory (the idea one can infer from sample to population, which is the cornerstone of null hypothesis testing) tends to downplay the fact that it is problematic to extrapolate from samples to populations in this regard.

Schmidt argued that only meta-analysis can get around these problems.* Again, meta-analysis is problematic. First, the publication bias still applies because published studies tend to use only significant findings [63]. Second, the Fisherian inference model is based around the idea that the NHST experiment is conducted *once and once only*. So, in other words, replication (surely one of the key aspects of scientific credibility) cannot really ever take place in a null hypothesis framework.

---

* Meta-analysis combines the result of independent experiments or studies. Results from larger studies can be given more weight. Two basic models exist. The fixed-effects model assumes random variation in individual studies based around a single underlying effect (i.e., if all studies were large enough, they would report the same effect). The random-effects model assumes a different underlying effect for each study and takes this into consideration. They do not often lead to substantially different results. There are critiques of the meta-analytic approach that are beyond the scope of this text. It is sufficient to say that it is still rare to see meta-analytic studies in most areas of psychology or human factors.

(In fact, Bakan [64] argued that replication of an experiment actually destroys the validity of the traditional inferential process unless the replication is taken into account and *the probabilities revised accordingly.* It is hoped that this idea of revision of probabilities will bring readers to mind of the arguments about Bayesian inference and conditional probabilities in Chapter 3.)

To see why the classic Fisherian NHST presupposes one and only one experiment, consider the following example. Suppose we carry out a trial on the performance of two safety devices, and we reject the null hypothesis (no difference in performance) at $p < 0.05$. How are we to interpret this situation? If we are interpreting the test correctly, we will say that in fewer than 1/20 or 5% of trials we would get a sample of data like the ones gathered if it were true that there was no difference between the two devices.*

But (and it is a very big but), this ignores the *actual* results of other studies. Say there were 19 identical trials that we did not know about (carried out at different laboratories or whatever), and in all these *other* trials the *Ho* was *not* rejected. Then, this one significant result out of 20 (which we thought had proved our point) would be precisely what we would expect by chance when there was really no difference between the samples. The probability of a data set occurring given the *Ho* begins to sound fairly meaningless unless we take into account all trials that shed light on this.

4.3.6 Why Did the *Ho* Test Become So Commonplace?

The point we would make here is that psychology and safety science were both predisposed to become confused in this particular way. Cronbach [65] noted two disciplines in measuring human behavior, the experimental and the correlational (or probabilistic). Experimental psychology (and safety science) have long been associated with the seeking of general causal deterministic laws (see Chapter 1). Given that the laws sought are causal/deterministic (not statistical), if there is even a single example of the law not working, then the law must be discarded or modified. Unfortunately, the experimentalist, searching for Newtonian absolutes, stumbled across the correlationists' preferred statistical tool and decided to appropriate it for his or her own ends.

But, *Ho* tests, which epitomize the experimental approach, are underpinned by probabilistic logic** and therefore are wholly unsuitable for the demonstration (or otherwise) of general causal laws. Yet, Bakan cited unpublished data from Roberts and Wist, who examined 60 articles from psychological literature and found 25 clear cases for which general (law)-type conclusions had been drawn from aggregate type data [66]. The profound irony is that *Ho* testing, the preferred tool of the objectivist

---

* Another problem is that people routinely misinterpret the $p < 0.05$ result as meaning that in fewer than 1/20 or 5% of cases there will be a nonsignificant result, or that in 95% or more cases a significant difference between the devices will be found. It must be stressed that the results of a single trial say nothing about the likelihood of replication of a particular p value or the likelihood of someone else rejecting or failing to reject a particular *Ho*. This depends on power (see above) and effect size and cannot be inferred from the p value of a single trial.

** That is, they generate probabilistic conclusions about which there is a natural degree of uncertainty.

and saturated in the ideology of technical rationality, is totally useless for providing support for the existence of the causal laws that the objectivist seeks.

A final point is that, even if *Ho* is used and interpreted correctly, it has "little to recommend it" because "the ritual dichotomous reject-accept decision, however objective and administratively convenient, is not the way any science is done" [67]. What we generally do is pick up on previous ideas and see whether we can become more or less sure of them. Real science is not binary or digital but analog and fuzzy, and it always has been. Pretending that its true purpose is to make strict (digital) yes/no-type decisions is pseudoscientific. As Cohen said:

> The prevailing yes-no decision at the magic 0.05 level from a single research is a far cry from the use of informed judgment. Science simply doesn't work that way. A successful piece of research doesn't conclusively settle an issue, it just makes some theoretical proposition to some degree more likely. ... There is no ontological basis for dichotomous decision making in psychological inquiry. [68]

4.3.7 WHY HAS THE TEST PERSISTED?

Regarding why the test has persisted, there are two answers we can think of (well, two that do not involve impoliteness about our fellow psychologists and ergonomists who continue to believe in *Ho* testing). The first is pragmatic and the second theoretical.

First, it is the way people tend to be "brought up," and it gets results. It gets you published in journals, it pleases regulators looking for test results or safety cases, and it underpins almost all theory put forward at conferences about the way the safety and risk world is. The pragmatic reasons for continuing down the *Ho* road were outlined rather bluntly by Meehl:

> Since the null hypothesis refutation racket is "steady work" and has the merits of an automated research grinding device, scholars who are pardonably devoted to making more money and keeping their jobs ... are unlikely to contemplate with equanimity a criticism that says that their whole procedure is scientifically feckless and that they should quit doing it and do something else. ... That might ... mean that they should quit the academy and make an honest living selling shoes. [69]

Schmidt notes that there may also be a *psychological* dependence on NHST (this may or may not be a joke): "The psychology of addiction to significance testing would be a fascinating research area" [70].

Second, the search for objectivity continues to rear its ugly head. Bakan [71] talked of the "dream, fantasy, or ideal" of what he called automaticity of inference. This refers to the idea that scientific inference can be objectified. In this dream world, instead of the responsibility resting on the scientist, who could be wrong and who would have to say "this is my opinion," the test of significance itself would, as it were, decide whether the hypothesis was true. This would allow a rigorous kind of operationalized inference that could be put across simply by reporting the statistical findings. Once more, we get the sense of the scientist motivated by the search for the Holy Grail of pure objectivity.

But, we note an arbitrary (subjective) probability level has to be set below which results will be significant. So, there is an element of subjectivity (arbitrariness) built into the project right from the beginning. And, the purpose of the test is to see whether the experiment, as it were, crosses that arbitrary line. Presumably worried about this subjective aspect to the setting up of the test, many experimenters have resorted simply to reporting p values as results, that is, failing to make a strict decision on significance (which is the *only* point of the test). This type of reporting (e.g., reporting simply that $p = 0.06$) is motivated by the implication that one has almost achieved a statistically significant result but not quite. This gives a kind of pseudo-objective measure of degree of confidence (this is where our old friend Thomas Bayes comes in again) by which people are seduced into comparison of individual p values rather than sticking to strict reject/fail to reject decisions required by the test.

Once more we are indebted to David Bakan for a succinct summing of why this is flawed:

> To regard the methods of statistical inference as exhaustive of the inductive inferences called for in experimentation is completely confounding. When the test of significance has been run, the necessity for induction has hardly been completely satisfied. However, the research worker knows this, in some sense, and proceeds, as he should, to make further inductive inferences. He is, however, still ensnarled in his test of significance and the presumption that it is the whole of his inductive activity, and thus mistakenly takes a low p value for the measure of the validity of his other inductions. [72]

So, if the test of significance is of limited usefulness and if we do not want to claim that some ultimate or absolute probability can be obtained, then how should we proceed? As in Chapter 3, we avoid discussing formal Bayesian statistics but show simple operations that can be performed on data sets using common statistical packages.

4.4 HOT SCIENCE

The alternative to the objectivist-experimental significance testing approach is broadly a nonexperimental approach. Bradley and Schaefer summed up the difference between the classical and nonexperimental approaches:

> Classical statistics, using experimental data, yields the likelihood that we have correctly accepted or rejected a null hypothesis, but with non-experimental data the null hypothesis itself has a "probability" or degree-of-belief attached to it. Probabilities are indexes of the reasonableness-of-doubt that should be attached to conclusions — conclusions that, as in court cases, generally cannot be proved or disproved but can be argued more or less persuasively. [73]

We believe the ultimate basis for any decisions we might make must be pragmatic and should be based on our experiences and the data we have observed. Plain old descriptive statistics are frequently illuminating [74]. However, using exploratory statistical procedures to test models (i.e., assumptions we may have about the world) rather than to test a null hypothesis provides a paradigm for ensuring we obtain the best value from data analysis. Tukey, who encouraged exploration of data sets, called this an emphasis on "detective work" rather than "sanctification" [75].

Desrosieres [76] outlined how the frequentist/experimental paradigm in science, which entails a search for general laws and a description therefore of general reality, can be contrasted with the Bayesian/epistemic approach, which he called a "science of clues." The latter is not concerned with general reality but rather with attempts "to establish local and specific chains of causation in order to explain particular events" [77]. The point is that the posterior probability simply provides better grounds for a specific action than the prior probability. Desrosieres pointed out that, in this science-in-the-making (or hot science), truth is still a guess, a bet. Only later, when science cools down again, are certain results encapsulated, becoming recognized facts, while others disappear altogether [78].

We want to argue safety management is hot science. Safety managers and engineers (even psychologists or ergonomists) are hot scientists, a description to which we hope most of our readers will not object strongly. The task is to make the best guess based on the information available, no more and no less. The role of statistical procedures should be *exploratory* rather than definitive with respect to a single unlikely hypothesis. So, this is *nonexperimental* in the classical sense. (Again, this should remind us of the distinction between Newtonian and Goethian science or between Platonic and Aristotelian science as discussed in Chapter 1. The exploratory, hot science of clues is clearly Goethian/Aristotelian [in the broadest sense] in the way that the opposing methodology is Newtonian/Platonic [again, in the broadest sense].)

In the next section, we briefly look at descriptive statistics for classificatory data before discussing slightly more intricate analysis of the types of categorical (taxonomic) data we introduced in the Chapter 3; we use the examples of a cluster analysis and a log linear analysis (with exploration as a goal [79]). Readers can assume several things prior to this section: there will be no use of the null hypothesis, no making of unwarranted inductive leaps from probabilities relating to hypothetical trials, and no claim that objective probabilities can be or have been obtained.

4.5 WORKING WITH TAXONOMIC DATA

Let us backtrack for a second. We have argued that all data (whether they originate from analysis of texts, videos, or audiotapes; direct observation of events; technical measurement of physical or environmental phenomena) are *taxonomic*. They are simply classified pieces of information. Let us assume we are looking at classificatory data of the type we looked at in Chapter 3, and we have a tightly organized database underpinned by reliability trials. What can we *do* with these data without falling into some of the objectivist and experimentalist traps discussed? The rest of this chapter shows some techniques and analyses that are possible.

It is important to stress that we are, as in Chapter 3, counterpointing Bayesian with frequentist ideas on probability and suggesting that the dominant frequentist paradigm is often used inappropriately. But, none of the techniques we mention briefly here are, strictly speaking, Bayesian (although examples like cluster analysis and log linear analysis have been described by Byrne as broadly compatible with subjective probability) [80]. The reason we are not going to discuss genuine Bayesian analysis is that Bayesian software is still relatively hard to obtain and not particularly user friendly (although this situation is improving).

The real aim of this chapter is to dispel a myth. Many psychologists, safety managers, and ergonomists have been brought up on null hypothesis/frequentist suppositions and assume that this is the only way to do real science. And, if it is suggested (as it is here) that these techniques are best abandoned, this can lead to a certain metaphysical panic. Frequently, it is assumed that there is no alternative to frequentist statistics and that if they are abandoned we may as well go back to using astrological or alchemical techniques and disregarding Western science altogether. This is nonsense. There are many coherent exploratory and descriptive techniques that produce reliable and meaningful data. The following list is not in any sense meant to be exhaustive but simply to function as a taste of what is possible when the search for objective (frequentist) probabilities is abandoned.

4.5.1 FREQUENCY COUNTS

First, we can simply count the data elements coded into each category, such as events on a Tuesday, communication problems between pilots and copilots, boiler shutdowns, the number of times temperature parameters have been exceeded, and so on. This is called frequency analysis or bean counting. Such an analysis will often be presented in the form of a bar chart or pie chart in which individual bars or sections refer to each category. We hope readers are familiar with such counts and charts. This simple form of analysis can be useful in that it may illuminate certain problem areas and lead to meaningful intervention strategies.

We would sound two notes of caution. At the risk of sounding like a broken record, the usefulness of such analysis depends on the reliability of classification. Any conclusion drawn will be invalid if events are not classified into the same (or similar) categories by different safety personnel. Remember, we want the bean counts to reflect different behavior in the event reports themselves, not differences in behavior at the operational feedback desk.

Second, comparison should be between categories at a *particular level* in our taxonomic organization (see Figure 3.1 for an example of a hierarchical taxonomy). Otherwise, high frequencies (e.g., high bars in our bar chart or big slices in our pie charts) will simply reflect differences in the conceptual size of the categories. We do not want a bar chart that shows technical problem frequencies against those for procedure omitted information. The former will invariably show up higher because it is a much broader box (it would usually split into further categories). We should be concerned with comparisons of *types of technical problem* or *types of procedural problem*, for which different frequencies will be meaningful in discriminating between events rather than merely indicating the conceptual size of the codes (or the level in our hierarchy at which the code appears).

4.5.1.1 Trending and Control

It is also the case that, sometimes, bean counting can be misleading. There may be a special reason for a particular spike that we need to take into account before using it to inform decision making. The principle of trend analysis and time series analysis is that we look at an ordered sequence of values of something at equally spaced time intervals, so that we can tell whether our bean counts reflect a trend (a predictably shifting pattern) or are best treated simply as snapshots. It is perfectly feasible to look at frequencies of output from categorical or taxonomic decisions over time. The simple idea is to see whether something is falling, rising, or remaining static.

A simple line chart (like a bar chart, but instead of different categories, we compare frequencies for one category at different time points) can tell us what may be happening with a category of event or a particular human factor over time. However, such a chart can sometimes resemble the Karakorum mountain range, with a sequence of high jaggy peaks and troughs, which can be hard on the eye (and brain); it may be desirable to smooth the data to aid interpretation. A range of smoothing techniques is available, the description of which is beyond the scope of this text. An example, however, would be the computation of a simple moving average. This involves looking at a set of frequencies and computing the mean frequency of consecutive smaller sets. Suppose we examine the frequency of scheduling issues over time and get a sequence of monthly levels of this that reads 9, 8, 9, 12, 9, and 12. It is relatively difficult to make sense of these data regarding whether a trend is emerging. (This process can become fairly sophisticated; 50- and 200-day moving averages are favorite tools of share analysts.)

Now, say we compute the mean of the first three numbers; we obtain $26/3 = 8.67$. We can advance this averaging process one number at a time. So, the next mean frequency is for the second, third, and fourth numbers $= 29/3 = 9.67$. The next mean is 10, and the one after that is 11. Now, we have a sequence of 8.67, 9.67, 10, and 11, which is easier to interpret (it indicates a steady rise in frequencies).

We do not go into trending techniques in any more detail here. Techniques are available to try to obtain further understanding of trending data. It should be remembered that when underlying distributions are not taken into account, rises in actual event frequencies like the sequence above may be spurious. The principle of control charts is to take normal variance into account and to flag unusual variance. In previous research, we have created such charts, in which parameters are set on various variables; if these variables exceed these parameters, then a red flag is waved indicating the need for further investigation. Human factors variables are merely variables like any other, and we have created such control charts for human factors data from, for example, data from the railway industry. Bayesian methodologies for control charts are increasingly available [81].

4.5.1.2 Cross Tabulation

Another useful technique with taxonomic data is to look for *associations* between categories by means of cross tabulation. Although the bean counts give a flavor of the types of factors that appear regularly in our database, cross tabulation simply involves looking at one factor for various levels of another.

For example, we may have a bean count that shows many slips and trips. It is reasonable as a next exploratory step to look at *where* or *when* slips and trips occur (i.e., in the language of taxonomic theory, to *condition* the original count on *some other taxonomic decision*). If we found out that slips and trips were occurring with regularity in a certain corridor, then this would lead us to think that maybe there was something specific about that corridor that we should be looking at (perhaps the floor is made of a particularly slippery substance or whatever).

Obviously, we can cross tabulate any two or more factors for which we have reliable taxonomies: slips and trips at particular times of day, communication problems at different plants, fatigue issues under different supervisors, and so on. The usefulness of this is fairly obvious. The usual way of presenting cross-tabulated data is by a contingency table, with rows and columns representing coding categories and the frequencies in the boxes or cells indicating codes assigned. (An example is shown in Table 3.8 in Chapter 3, which is in effect a cross tabulation that looks at associations between two coders. Exactly the same format is used to show cross tabulation and association between two factors or codes.)

It should be mentioned that such cross tabulation is traditionally followed by a test of statistical significance (usually a χ^2 test), which will give a *p* value to allow us to decide whether to reject the null hypothesis that there is absolutely no association between our categories. In Section 4.3, we outlined why we do not think this is a good use of time and effort. If we are investigating communication problems on weekends versus weekdays, we clearly already think there might be an association; otherwise, we are strange safety managers investigating relationships between things we think are unrelated. Cross tabulations, for the graphically minded, can also be presented as bar charts, such as using different colored bars side by side or stacked on top of one another to indicate the presence or absence of one code within the overall frequency of another.

4.5.2 Cluster Analysis

Another technique that avoids the use of frequentist statistical inference (from sample to population) is cluster analysis. It is not the purpose of this chapter to provide a definitive account of how to carry out a cluster analysis. Rather, the aim is to give a basic understanding of how cluster analysis works and how useful it might be in a safety context. Everitt provided a good introduction to the methodology and a comparison of the various techniques [82]. (Cluster analysis can be easily carried out using the computer package *SPSS* [*Statistical Package for the Social Sciences*] or most other statistical packages.)

Cluster analysis, as the name suggests, is simply a way of organizing data such that clusters, groups, or categories emerge from the analysis. Now, readers may at this point feel a vague sense of recognition of the term *categories* and question whether this sounds like a cluster analysis produces a taxonomy of sorts? The answer is, quite clearly, yes. What emerges from a cluster analysis of safety events is what most authors would call a numerical taxonomy [83, 84].

The type of taxonomy discussed in this book to this point is generally called a typological taxonomy. Bailey [85] would say our coding taxonomies (see Chapter 3) are conceptual and the taxonomy arising from a cluster analysis of events is empirical. But, it has been pointed out that, in practice, there is much overlap. Coding taxonomies of the type discussed in Chapter 3 are most often designed through empirical processes (like observation of events, interviews with staff), and cluster analysis techniques generate results that are partly defined by the concepts we bring to the process (e.g., there is no pure induction; we always have to decide which data we choose to use to examine events) [86].

In simple terms, we have argued in the previous chapters that classifying safety events to discriminate between them is important. What we have after such a process is a reliable classification of each event using a series of categories. What we are going to talk about here is using these classifications to say something about groups of safety events in general terms.

Traditionally, categories used to discriminate between cases or events would be called *variables*, which are assumed to be something real that can be measured. A traditional variable-centered approach concentrates on interactions between these features, which are seen as discrete entities (i.e., objective features of the world separate from the events we are studying). The mechanism arises out of (or is inferred from) a causal analysis in which a relationship between these variables is established. This approach is related to the search for general causal laws we mentioned in Chapter 1.

However, Byrne argued strongly against this idea of a variable that is reified and abstracted from systems or cases. What he said we do when we classify events (see Chapter 3) is that we measure traces of systems that are both incomplete and dynamic. As he put it: "The dynamic systems which are our cases leave traces for us, which we can pick up" [87]. He argued that exploring systems by means of classifications we have made maintains an Aristotelian "engagement with the world as it is."

The general principle that Byrne outlined and with which we agree is that, if the world is complex in the technical sense of that word (and it surely is), then it cannot be understood by inferring causal relationships between abstract and supposedly static variables. We have argued elsewhere that measurement (classification of traces of systems) is a *hermeneutic* process [88], something with which Byrne concurred when he said, "Given a generally exploratory view of measurement, then we can see measurements as provisional versions of reality which can be improved by re-engagement with the account they give."

In a cluster analysis, events are grouped together (*clustered*) on the basis of things we know about them. Predictions can then be made about a small number of groups of events (named *clusters*) rather than a very large number of individual events. So, the trick in cluster analysis is to collect information and combine it in ways that allow classification of events into useful groups. Information, crucially, can be about human factors as well as other aspects of a system and can be in binary form (i.e., present/absent codes of the type discussed).

Cluster analysis can be contrasted to the more widely used factor analysis, which treats variables as indeed having separate properties. Factor analysis groups these objective variables together to create more general ones (factors). Cluster analysis groups events or cases together instead, treating the variables as dynamic properties of the cases that are used to place a group of cases close to each other (i.e., cluster them) because they have similar dynamic properties or separate them if they do not.*

Finally, cluster analysis techniques do not sort events into preexisting categories. The clusters that emerge are not known beforehand and cannot be inferred from knowledge of the event traces. This idea of emergence is a key theme of this book and an important feature of exploratory (rather than test-based or variable-based) data analysis.

Predictably, frequentist statisticians do not like clustering techniques, essentially because "they are not constructed around a central concern with inferences from samples to universes" and there is an "absence of anything resembling *tests of significance* in the most commonly used clustering procedures" [89; italics added] and because subjective decisions (which classifications to make, which clustering method or algorithm to use) lead to different clusters or taxonomies of cases. Readers should know by now what we think of significance tests and the idea that scientific inquiry can somehow be "not subjective." Byrne agreed with us and said the following:

> Meaningful emergent classifications … are robust. Actually it seems to me that this property of robustness is a better source of validation than statistical inference because it more closely resembles the idea of repeatability which, supposedly, underpins the validation of experimental evidence. Basically, if there is a real underlying taxonomy to be found then different clustering methods will produce essentially similar classification. [90]

(One might argue that this is similar to our view of reliability; that is, different people analyze the same data, and then we see the degree of consensus between their observations.)

We take it as axiomatic that we want to reduce unwanted events. There are various useful (and less-useful) pieces of information about the events and people involved in them that might allow for prediction and therefore action. However, it is usually not practical to go through all possible evidence for all events and to make separate predictions based on each piece. What a cluster analysis does, in essence, is help avoid the need for such legwork by grouping events together. More general predictions can then be made about event types.

To summarize, in a cluster analysis, there are no *p* values or absolute probabilities. Yet, what we do is statistical. The taxonomy of events we produce is based on reliable data about real events. If events cluster, then the taxonomy we are using is important for predicting events. A good example would be the clustering of minor events around generic human factors codes that would allow for targeted action.

We might finally note that cluster analysis is relatively uncommon in human factors and psychology (for which absolute probability and tests of significance abound), yet the opposite is true in marketing, for which cluster analyses are routine. For example, Larson [91] described how census information (e.g., population density, income, age) was used to define neighborhoods. The emergent clusters of

---

* In other words, factor analysis looks through the data to see alleged variables that lie behind or beyond the data, whereas cluster analysis looks at the data themselves. This is a fundamental difference and should remind the reader of the conduit metaphor of language in Chapter 1.

these neighborhoods were then named Blue Blood Estates, Shotguns and Pickups, Bohemian Mix, etc. This classification scheme (numerical taxonomy), called the Potential Rating Index for Zip Markets (PRIZM), has subsequently been used in direct mail advertising, radio station formats, and decisions about where to locate shops. As pragmatists at heart, we feel that such an endorsement from the money people tells its own story.

4.5.3 LOG LINEAR ANALYSIS

A final example of data analysis we would like to mention is log linear analysis. Byrne explained that the technique is "based on a comparison between the actual observed data ... and that which is generated by a mathematical equation which describes the model being compared — note the avoidance of the word tested" [92].

So, we try to construct a model (if you like, a subjective set of beliefs we have about the system) that accounts for or explains the majority of the observed data (observed data, readers will appreciate, we see as frequencies generated by reliable taxonomic classification). This is generally known as goodness-of-fit testing.

Gilbert pointed out the clear differences between this approach and the classical one involving the generation of a null hypothesis:

> Statistical texts ... generally consider the testing of hypotheses rather than models. ... While a hypothesis usually concerns just one relationship, a model may and usually does involve a complex set of linked relationships. Secondly, the classical approach assumes that the analyst possesses, *a priori*, a carefully formulated hypothesis to be tested against the data. Following the confirmation or rejection of this hypothesis, the analyst must cease working with the original set of data. Improved hypotheses should be tested with new data. In contrast [the exploratory approach] assumes that ... the investigation ceases only when an adequate model to describe one set of data has been found. The task is essentially to explore in depth the structure of the data. This seems a more realistic view of research in practice than the classical one. [93]

Significance tests in log linear analysis simply tell us how close our model is to the data. This is not like a classical significance test (like a χ^2 test) in which the observed data are compared with a hypothetical frequency under a null hypothesis. In other words, a log linear analysis measures for *closeness* (i.e., similarity, goodness of fit) rather than *difference*. The Bayesian parallels were outlined by Bradley and Schaefer:

> Probabilities here *are degrees of warranted belief.* ... Probability is not a statement about physical events but an estimate of the level of believability, the relative weight of admissible evidence in the face of uncertainty and ignorance; it is an assessment of the likelihood of a particular conclusion, given a body of (imperfect) information. [94; italics added]

So, we classify events, then we look at data we have generated. The actual data are accounted for by complex relationships between all the classifications we have made. But, we hope to say something general about the system using a model that is simpler than the actual model (involving all relationships) yet does not lose too much useful detail. The simple idea of a log linear analysis is to reduce the complexity of a total or saturated model (one that specifies *every single* relationship in a data set) while losing as little *predictive value* as possible.

With a log linear analysis, no hypotheses are rejected or accepted; we simply end up with a simplified description (model) of the data set with which we are working. This typifies the exploratory approach.

4.6 CONCLUSION

The important thing about the types of analysis we are proposing is that they deal with patterns, which is vital when we deal with systems, for which (as we discuss in later chapters) linear causality does not apply. The conclusions we have shown in our examples deal with patterns, organizations, and interactions. So, our unit of analysis here is the totality of the dynamic system features instead of the individual atomized elements that make up the system. In other words, instead of sampling from the system (and then using null hypothesis testing to generalize hypotheses from this sample to the whole model), we look at the *entirety* of the model/system. Therefore, this approach is not about inferences from samples, but about modeling data sets, saying something about the data we have gathered, and using findings to drive action and inform our beliefs about the systems with which we are working. We produce *models of systems* rather than make decisions about very strict propositions involving individual variables.

We should note that cluster analysis does not so much test hypotheses as *generate* them. Having clustered events, we should now have some idea about how to proceed. We now believe certain things about our system. If we address the issues we have seen to be important in grouping events, then we can see how this affects the system. We have now some clear discrimination and, it is hoped, understanding about the dynamics of multiple events and multiple human factors.

We hope that, as we come to the end of two chapters on working with taxonomies, it will be clear why we agree with Byrne when he said, "Classification is perhaps the most important way in which we can understand a complex non-linear world" [95].*

The data we work with are real, as are the plants we manage and the technical items and people contained in them. We see no reason that the probabilities and concepts we use should be abstractions or should apply to hypothetical (or unlikely) scenarios. We prefer to condition our subjective beliefs on what we have found or seen rather than condition our data on (peculiar) hypotheses that only statisticians seem to understand or use correctly. Unfortunately, there is a long tradition in psychology of a concern with the abstract and the hypothetical, to which we now turn.

---

* Byrne outlined how a numerical taxonomy of events produced in this way is, of course, a further classification (which is where we began). So, the process can become reflexive, as we use cluster membership itself to input into, for example, a log linear model.

REFERENCES

1. Schön, D.A., *The Reflective Practitioner*, Basic Books, London, 1991.
2. Byrne, D., *Interpreting Quantitative Data,* Sage, Thousand Oaks, CA, 2002, pp. 7–8.
3. Tukey, J.W., *Exploratory Data Analysis,* Addison-Wesley, Reading, MA, 1977.
4. U.S. Nuclear Regulatory Commission, *Reactor Safety Study: An Assessment of Accident Risks in U.S. Commercial Nuclear Power Plants,* Government Printing Office, Washington, D.C., 1975.
5. Sugarman, R., Nuclear power and the public risk, *IEEE Spectrum,* 16, 11, 1979.
6. Cooke, R.M., *Experts in Uncertainty: Opinion and Subjective Probability in Science,* Oxford University Press, New York, 1991.
7. Miller, C.K., The presumptions of expertise: The role of ethos in risk analysis, *Configurations,*11(2), 183, 2003.
8. Risk Assessment Review Group, *Report to the U.S. Nuclear Regulatory Commission,* U.S. Nuclear Regulatory Commission, Washington, D.C., 1978.
9. Risk Assessment Review Group, *Report to the U.S. Nuclear Regulatory Commission,* U.S. Nuclear Regulatory Commission, Washington, D.C., 1978, p. 3.
10. von Hippel, F., Looking back on the Rasmussen report, *Bulletin of the Atomic Scientists,* 33(2), 44, 1977.
11. Miller, C.K., The presumptions of expertise: The role of ethos in risk analysis, *Configurations,* 11(2), 180, 2003.
12. Risk Assessment Review Group, *Report to the U.S. Nuclear Regulatory Commission,* U.S. Nuclear Regulatory Commission, Washington, D.C., 1978, pp. 9–10.
13. Abramson, L.R., The philosophical basis for the use of probabilities in safety assessments, *Reliability Engineering and System Safety,* 23, 255, 1988.
14. Bley, D., Kaplan, S., and Johnson, D., The strengths and limits of PSA: Where we stand, *Reliability Engineering and System Safety,* 38, 3, 1992.
15. Chibber, S., Apostolakis, G., and Okrent, D., A taxonomy of issues related to the use of expert judgments in probabilistic safety studies, *Reliability Engineering and System Safety,* 38, 27, 1992.
16. Hora, S.C., Acquisition of expert judgment: examples from risk assessment, *Journal of Energy Engineering,* 118, 136, 1992.
17. Aven, T. and Pörn, K., Expressing and interpreting the results of quantitative risk analyses. Review and discussion, *Reliability Engineering and System Safety,* 61(1), 3, 1998.
18. Redmill, F., Exploring subjectivity in hazard analysis, *Engineering Management Journal,* 12(3), 142, 2002.
19. Poucet, A., *HF-RBE Human Factors Reliability Benchmark Exercise — Synthesis Report* (EUR12222 EN), Commission of the European Communities, Luxembourg, 1989.
20. Embrey, D.E., Humphreys, P., Rosa, E.A., Kirwan, B., and Rea, K., *SLIM-MAUD: An Approach to Assessing Human Error Probabilities Using Structured Expert Judgement* (NUREG/CR-3518), BNL, New York, 1984.
21. Swain, A.D. and Guttmann, H.E., *Handbook of Human Reliability Analysis with Emphasis on Nuclear Power Plant Application* (NUREG/CR-1278), India Labs, Albuquerque, NM, 1980.
22. Hollnagel, E., *Cognitive Reliability and Error Analysis Method (CREAM),* Elsevier Science, Oxford, U.K., 1998, p. 132.
23. Apostolakis, G., The concept of probability in safety assessments of technological systems, *Science,* 250(4986), 1361, 1990.

24. Redmill, F., Subjectivity in hazard analysis, *Journal of System Safety*, 40(1), 2004 (accessed online at www.system-safety.org/index).

25. Apostolakis, G., The concept of probability in safety assessments of technological systems, *Science*, 250, 1360, 1990.

26. Hollnagel, E., *Cognitive Reliability and Error Analysis Method (CREAM)*, Elsevier Science, Oxford, U.K., 1998, p. 122.

27. Redmill, F., Risk analysis: A subjective process, *Journal of System Safety*, 39(2), 2003 (accessed online at www.system-safety.org/index).

28. Desrosieres, A., *The Politics of Large Numbers: A History of Statistical Reasoning*, Harvard University Press, Cambridge, MA, 1998.

29. Byrne, D., *Interpreting Quantitative Data*, Sage, Thousand Oaks, CA, 2002, pp. 7–8.

30. Byrne, D., *Interpreting Quantitative Data*, Sage, Thousand Oaks, CA, 2002, p. 80.

31. Edwards, W., Lindman, H., and Savage, L.J., Bayesian statistical inference for psychological research, *Psychological Review*, 70(3), 213, 1963.

32. Cohen, J., The earth is round ($p < .05$), *American Psychologist*, 49(12), 997, 1994.

33. Bakan, D., *On Method*, Jossey-Bass, San Francisco, 1967.

34. Meehl, P.E., Theory testing in psychology and physics: a methodological paradox, *Philosophy of Science*, 34, 103, 1967.

35. Rozeboom, W.W., The fallacy of the null hypothesis significance test, *Psychological Bulletin*, 57(5), 416, 1960.

36. Rozeboom, W.W., The fallacy of the null hypothesis significance test, *Psychological Bulletin*, 57(5), 416, 1960.

37. Cohen, J., The earth is round ($p < .05$), *American Psychologist*, 49(12), 998, 1994.

38. Cohen, J., The earth is round ($p < .05$), *American Psychologist*, 49(12), 997, 1994.

39. Cohen, J., The earth is round ($p < .05$), *American Psychologist*, 49(12), 997, 1994.

40. Gigerenzer, G., The Superego, the Ego, and the Id in statistical reasoning, in Keren, G., and Lewis, C. (Eds.), *A Handbook for Data Analysis in the Behavioral Sciences: Methodological Issues*, Erlbaum, Hillsdale, NJ, 1993, p. 311.

41. Byrne, D., *Interpreting Quantitative Data*, Sage, Thousand Oaks, CA, 2002, p. 81.

42. Matloff, N.S., Statistical hypothesis testing: problems and alternatives, *Environmental Entomology*, 20(5), 1246, 1991.

43. Gardner, M.J. and Altman, D.G., Confidence intervals rather than *p* values: estimation rather than hypothesis testing, *British Medical Journal*, 292, 746, 1986.

44. Zeisel, H., The significance of insignificant differences, *Public Opinion Quarterly*, 17, 319, 1955.

45. Cohen, J., The earth is round ($p < .05$), *American Psychologist*, 49(12), 998, 1994.

46. Cohen, J., The earth is round ($p < .05$), *American Psychologist*, 49(12), 1000, 1994.

47. Edwards, W., Lindman, H., and Savage, L.J., Bayesian statistical inference for psychological research, *Psychological Review*, 70(3), 214, 1963.

48. Bakan, D., *On Method*, Jossey-Bass, San Francisco, 1967, p. 7.

49. Nunnally, J.C., The place of statistics in psychology, *Educational and Psychological Measurement*, 20, 641, 1960.

50. Popper, K., *The Logic of Scientific Discovery*, Hutchinson, London, 1959.

51. Cohen, J., The earth is round ($p < .05$), *American Psychologist*, 49(12), 999, 1994.

52. Meehl, P.E., Theory testing in psychology and physics: a methodological paradox, *Philosophy of Science*, 34, 103, 1967.

53. Thompson, B., Two and one-half decades of leadership in measurement and evaluation, *Journal of Counseling and Development*, 70, 436, 1992.

54. Bakan, D., *On Method*, Jossey-Bass, San Francisco, 1967, p. 17.

55. Cohen, J., The earth is round ($p < .05$), *American Psychologist*, 49(12), 997, 1994.

56. Cohen, J., *Statistical Power Analysis for the Behavioral Sciences,* 2nd ed., Erlbaum, Hillsdale, NJ, 1988.

57. Wilkinson, L. and APA Task Force on Statistical Inference, Statistical methods in psychology journals: Guidelines and explanations, *American Psychologist,* 54(8), 594, 1999.

58. Vacha-Haase, T., Nilsson, J.E., Reetz, D.R., Lance, T.S., and Thompson, B., Reporting practices and APA editorial policies regarding statistical significance and effect size, *Theory and Psychology,* 10(3), 413, 2000.

59. Thompson, B., Statistical significance and effect size reporting: Portrait of a possible future, *Research in the Schools,* 5(2), 33, 1998.

60. McLean, J.E. and Ernest, J.M., The role of statistical significance testing in educational research, *Research in the Schools,* 5(2), 15, 1998.

61. Sterling, T.D., Publication decisions and their possible effects on inferences drawn from tests of significance — or vice versa, *Journal of the American Statistical Association,* 54, 30–34, 1959.

62. Schmidt, F.L., What do data really mean? Research findings, meta analysis, and cumulative knowledge in psychology, *American Psychologist,* 47(10), 1173, 1992.

63. Egger, M. and Smith, D., Meta analysis bias in location and selection of studies, *British Medical Journal,* 316, 61, 1998.

64. Bakan, D., *On Method,* Jossey-Bass, San Francisco, 1967.

65. Cronbach, L.J., The two disciplines of scientific psychology, *American Psychologist,* 12, 671, 1957.

66. Roberts, C.L. and Wist, E., *An Experimental Sidelight on the Aggregate General Distinction* (unpublished manuscript, cited in Bakan, D., *On Method,* Jossey-Bass, San Francisco, 1967, p. 23).

67. Cohen, J., The earth is round ($p < .05$), *American Psychologist,* 49(12), 999, 1994.

68. Cohen, J., Things I have learned (so far), *American Psychologist,* 45(12), 1304, 1990.

69. Meehl, P.E., Theory testing in psychology and physics: a methodological paradox, *Philosophy of Science,* 34, 103, 1967.

70. Schmidt, F.L., What do data really mean? Research findings, meta analysis, and cumulative knowledge in psychology, *American Psychologist,* 47(10), 1176, 1992.

71. Bakan, D., *On Method,* Jossey-Bass, San Francisco, 1967, p. 16.

72. Bakan, D., *On Method,* Jossey-Bass, San Francisco, 1967, p. 21.

73. Bradley, W.J., and Schaefer, K.C., *The Uses and Misuses of Data and Models: The Mathematization of the Human Sciences*, Sage, London, 1998, p. 148.

74. Tukey, J.W., The future of data analysis, *Annals of Mathematical Statistics,* 33, 1, 1962.

75. Tukey, J.W., The future of data analysis, *Annals of Mathematical Statistics,* 33, 1, 1962.

76. Desrosieres, A., *The Politics of Large Numbers,* Harvard University Press, Cambridge, MA, 1998.

77. Byrne, D., *Interpreting Quantitative Data,* Sage, Thousand Oaks, CA, 2002, p. 81.

78. Desrosieres, A., *The Politics of Large Numbers,* Harvard University Press, Cambridge, MA, 1998, p. 5.

79. Byrne, D., *Interpreting Quantitative Data,* Sage, Thousand Oaks, CA, 2002, chap. 6.

80. Byrne, D., *Interpreting Quantitative Data,* Sage, Thousand Oaks, CA, 2002, chap. 5.

81. Bayarri, M.J. and Garcia-Donato, G., A Bayesian sequential look at u-control charts, *Technometrics,* 47(2), 142, 2005.

82. Everitt, B., *Cluster Analysis*, Wiley, New York, 1993.

83. Bailey, K.D., *Typologies and Taxonomies: Quantitative Applications in the Social Sciences,* Sage, London, 1994, p. 102.
84. Byrne, D., *Interpreting Quantitative Data,* Sage, Thousand Oaks, CA, 2002.
85. Bailey, K.D., *Typologies and Taxonomies: Quantitative Applications in the Social Sciences,* Sage, London, 1994, p. 102.
86. Byrne, D., *Interpreting Quantitative Data,* Sage, Thousand Oaks, CA, 2002, p. 100.
87. Byrne, D., *Interpreting Quantitative Data,* Sage, Thousand Oaks, CA, 2002, p. 36.
88. Wallace, B., Ross, A.J., and Davies, J.B., Applied hermeneutics and qualitative safety data: The CIRAS project, *Human Relations,* 56(5), 587, 2003.
89. Byrne, D., *Interpreting Quantitative Data,* Sage, Thousand Oaks, CA, 2002, p. 100.
90. Byrne, D., *Interpreting Quantitative Data,* Sage, Thousand Oaks, CA, 2002, p. 100.
91. Larson, E., *The Naked Consumer,* Holt, New York, 1992.
92. Byrne, D., *Interpreting Quantitative Data,* Sage, Thousand Oaks, CA, 2002, p. 91.
93. Gilbert, N., *Analyzing Tabular Data,* UCL Press, London, 1993, p. 6.
94. Bradley, W.J. and Schaefer, K.C., *The Uses and Misuses of Data and Models: The Mathematization of the Human Sciences*, Sage, London, 1998, p. 148.
95. Byrne, D., *Interpreting Quantitative Data,* Sage, Thousand Oaks, CA, 2002, pp. 127–128.

5 Psychology and Human Factors

5.1 TAXONOMIES AND PSYCHOLOGY

In the first four chapters, we looked at the basics of what we termed taxonomy theory, which is our approach to processing and analyzing data gained from almost any kind of situation, particularly safety situations. We hope the field is of interest in itself, but perhaps not surprisingly, working in the area has prompted us to examine the psychological theories commonly used in safety science and psychological theory in general.

There were two reasons for this. First, our approach to taxonomies presupposed that taxonomies or categories are *not* "hard wired" in the human brain from birth (if this was the case, then there could be, for all intents and purposes, taxonomies that were objectively true). Not all psychologists accept this, so we had to find out why they did not and whether their arguments held up.

The second reason is that it is noticeable that, if one looks at contemporary taxonomies (especially those in the field of cognitive ergonomics and the like), taxonomic descriptions of error and accidents are increasingly concerned with cognitive models or cognitive modules, frequently as error-causing events [1–4]. However, the introduction of psychological terminology such as this generally has not increased the reliability (or validity) of taxonomies currently in use (even granted that such reliability exists; the absence of adequate reliability tests on the vast majority of error taxonomies is something of a scandal in the field [5]). On the contrary, if anything, it seems to have made them *less* reliable via the introduction of unverifiable "inner states" (representations, schemas, etc.) as causal factors. This has made the classification of error even more difficult and has led to the field becoming even more jargon filled and imprecise than it was before.

How did this situation arise, and what can be done about it? We are *not* going to argue in this chapter that psychology currently has little to say to the safety practitioner because there is something fundamentally wrong with psychology. On the contrary, we argue that, when used properly, psychology can be useful indeed (we give examples of this). Instead, we argue that psychology seems to be of little practical relevance to current safety practice because of decisions that were made in the 1950s and 1960s regarding where psychology was going and what it ought to be. It was then that psychology abandoned the previously dominant behaviorist approach and instead adopted a new philosophy: cognitivism.

5.2 THE HISTORY OF COGNITIVISM

So, what is cognitivism, and why was it accepted so quickly and (it now seems) so uncritically? To begin, cognitivism is a tradition with many antecedents. If one wished to trace its ancestry back far enough, you could argue that its intellectual roots lie in the philosophy of Plato (just as with technical rationality [TR], and we argue later that this is not a coincidence). Certainly, René Descartes, John Locke, and other philosophers helped pave the way (again, in the same way as they paved the way for TR).

However, in the 20th century, cognitivism derives specifically from research initiated by Herbert Simon and Alan Newell in the then-new science of artificial intelligence in the 1950s and from work in linguistics carried out by Noam Chomsky in the 1950s and 1960s. The key point of cognitivism is that it viewed cognition as a matter of the manipulation of *symbols* by *rules*, rules that were conceptualized as similar to the algorithms used in formal logic. The symbols hypothesized by cognitivists were conceptualized as discrete (i.e., not fuzzy; they had hard edges like shapes in geometry) and digital (i.e., they either existed or they did not; they were not, in other words, qualitative or analog). These symbols existed in an internal cognitive space. Symbols were argued to be of two kinds: pure symbols (i.e., thoughts) and symbols of external reality (representations). However, both of these were manipulated in the same way: by algorithmic rules. Another key assumption of cognitivism was that one can make a hard-and-fast distinction between the brain and the mind in precisely the same way one can make a distinction between hardware and software on a personal computer. In other words, cognitivism was the view that the brain is a digital computer [6].

Now, to repeat what we have stated: this book is not a treatise on philosophy. Nevertheless, to see how this chapter fits in with the rest of the book, it has to be seen that cognitivism is, as it were, the counterpoint of the view of TR we examined in Chapter 1. TR looked at external reality as the study of discrete physical objects and the causal laws that described their movement. However, this led to an obvious problem: the mathematization of reality that this implied (such causal laws were best described via mathematical models) did not really seem to work when applied to human beings. Human beings, notoriously, do not seem to be predictable in the same sense that, for example, the movements of planets are. Therefore, at the beginning of the scientific revolution the now-notorious Cartesian dualism was created as a solution to this problem. This divided reality into two spheres: the external one of external physical reality and the internal one of consciousness and cognition. The methodology for studying external reality was proclaimed to be TR, but because the mind did not seem to be amenable to this form of analysis, it was argued that the mind was made of something different: mind stuff that was in some mysterious way radically different from the inert matter of the external world. Human beings were subjects; external reality was made of objects.* The outside world was subject to deterministic laws; our internal world was not.

---

* From this distinction, we get the distinction, so crucial to Western science, between subjectivity and objectivity. It is important to see that this binary dichotomy is not natural but is instead the result of a certain way of looking at knowledge and epistemology.

This approach continued until the mid-20th century when the cognitivists began to argue that this Cartesian dualism was fundamentally incoherent, but the solution they gave to the problem of Cartesianism was simply to ignore it. According to cognitivism, even the internal world of the mind is subject to deterministic laws. In other words, they coped with the problem of the distinction between the outer and inner by simply abolishing the inner (in the same way that they solved the problem of the subjective and the objective by simply abolishing the subjective). According to this view, we can study the mind in *precisely* the same way we study the external world: by breaking it down into its component parts or elements and studying the laws/rules that determine their movement. So, cognitivists looked at the internal rules that related the various cognitive elements (e.g., mental models) that made up our cognitive architecture. To be fair, instead of talking specifically about causal laws, they took over the language of stimulus–response from the behaviorists (which clearly descended from cause–effect, i.e., determinism) and substituted a middle third term to make stimulus–cognition–response. Nevertheless, the basic framework remained. The brain was now to be studied in precisely the same way as one studied the movement of the planets. The mind/body problem had been solved by abolishing the mind.

Here ends the philosophy lesson. But, the key point to remember is that cognitivism is the psychological wing of TR.*

5.2.1 OBJECTIONS TO COGNITIVISM

So, the essence of cognitivism is the idea that human cognition is law, rule, or algorithm bound (just like external reality), and that these rules are stored in the mind/brain. It has to be like this because (according to TR) the movement of objects in external reality was also rule (law) bound. Ergo, the internal world would also have to be described in the same way. However, this invites a rather important question: who uses these rules? To understand what this question means, imagine you are playing a game (say, chess), and your playing partner makes what you think is an illegal move. Who is it that picks up the rule book and looks up whether it is illegal? The answer is simple: *you* (or your partner) do. The same goes for any other kind of game; in terms of learning the game or determining whether a rule has been broken real human beings (the players, the umpire, the referee, or whoever) pick up, read, and interpret the rules. Without human beings, the rules would simply sit there as inert symbols in a book. It is easy to see that this is the same for any human activity: learning to drive, learning to program a computer, and so on. Certainly, there are rule books (for example the highway code in terms of learning to drive), but these need a human to pick up, read, interpret, and use the rules contained within them.

So, what about the rules that are, allegedly, inside your head? Who picks up and uses *them*? The answer, according to cognitivism, is that there is the equivalent of

---

* This is sometimes confused by the humanistic rhetoric with which cognitivism surrounded itself, especially in terms of the debate with behaviorism: as if comparison to a digital computer was less dehumanizing than comparison to a rat. The essence of the debate with behaviorism was that the cognitivists believed that behaviorism was not deterministic *enough*. Its laws (stimulus response, etc.) lacked the necessary deterministic character that the cognitivist sought.

a little man inside your head (a homunculus) who operates the rules. Sometimes, euphemisms for the homunculus are used (such as central processing unit or central executive), but the result is the same: there has to be some observer or user who activates, interprets, and uses the rules.

So, cognitivist theories are homuncular. However, this then leads to a paradox. By positing the existence of a homunculus, we merely beg the question: what is the motor of cognition of the homunculus? By the logic of cognitivism, the answer to this question must be ... more rules (the idea that *all* cognition is rule bound is one of the key assumptions of cognitivism). But, this then means we must posit *another* homunculus inside the homunculus to interpret and use *these* rules. And, assuming this homunculus also uses rules, then this homunculus in turn needs another homunculus inside *it*, and so on in a situation of infinite regress.

Of course, this has been noted as a problem, and attempts have been made to solve it. For example, the philosopher Daniel Dennett suggested that perhaps the homunculus has not one but two homunculi inside "his" head, and that these homunculi in turn have not two but four inside their heads, and so forth [7].* The difference between these homunculi is that every homunculus on the route down would be slightly stupider than the homunculi above them, with the chief homunculus the smartest of all. Eventually, we would get down to many, many, many little homunculi carrying out very simple tasks, which would at the very bottom simply be powered by neurons flicking on and off in a binary (digital) 0 and 1 fashion.

This answer is problematic in many ways (not least by asking, who breaks the tasks down and how?). However, the key problem is this: Dennett clearly means the on and off states of individual neurons to be equivalent to machine code, which is the basic binary language of digital computers (binary because it consists of ones and zeros). In other words, just as one breaks down a computer program into simpler and simpler programs until eventually one ends up with machine code, one can break down the program of cognition in the same way.

But, this misses an important point. The binary language of machine code *is itself a kind of rule* (algorithm, program). In other words, even after it has been broken down, it would still need a homunculus to interpret and use it. So, we are back in a situation of infinite regress again. In a computer, of course, the user is the homunculus, or to be more specific, human beings in general are the homunculi. Human beings build, program, and use digital computers. *We* provide the context within which the programs make sense, we make the programs, and we use them. The concept of cognition as rule bound is fundamentally incoherent.

There are many other arguments against cognitivism, but it is the homuncular argument that is the most devastating because it is completely irrefutable. It should be noted (contrary to the way this argument is sometimes interpreted) that the infinite

---

* Confusingly, Dennett might deny he was a cognitivist. This is because, as well as the antecedents mentioned above, cognitivist psychology derives from two different philosophical traditions, functionalism (Dennett was a functionalist) and representationalism (with its most famous exponent Jerry Fodor). Most psychologists tend to mix and match between these two traditions, although they are fundamentally incompatible. However, for our purposes it does not matter as functionalism (which posits similar internal mental architecture to representationalism, although these structures do not necessarily represent the real world) is just as prone to the homuncular fallacy as is representationalism.

regress argument is not simply a logical paradox that should (and can) be worked around. Instead, the argument shows that cognitivist explanations are not really explanations at all. What we are trying to explain is cognition, but what cognitivism does is simply push off that question by offloading the problem onto the homunculus.

However, it should be noted that the hypothesis that cognition is rule bound is not, strictly speaking, a scientific hypothesis. Instead, cognitivists tended to assume that because all behavior can be described as if it was the result of internal rules (which is of course true), therefore it was in fact the product of internal rules.* It is difficult to conceptualize scientific experiments that would prove or disprove this hypothesis; ultimately, it is a matter of faith. What we can do (via the homuncular argument) is to show that the whole concept is fundamentally incoherent.

However, we can do more than this. We can show that certain ancillary scientific hypotheses (e.g., the idea that memory is a store or that internal mental models exist), without which the rule-bound view of cognition collapses, are not in fact sustainable. Therefore, the rule-bound view of cognition collapses by default. So, it is to these other views we should now turn: again, first by asking if they are actually coherent views of cognition and then by looking at the relevant scientific evidence.

5.2.2 Symbols

According to the cognitivists, rules operated on symbols. We have seen that rule-bound cognition could not take place, but is it possible that a symbolic explanation of cognition could still be valid?

We take one specific kind of symbol first: a mental model. A mental model, it is argued, is a pictorial or quasi-pictorial model (or representation) of the external world. How do human beings see, act, and move? According to cognitivism, it was because we all have a mental model (or mental models) of the external environment. When we close our eyes, why are we not surprised when we open them to see the same view in front of us? Because, we were told, we have a mental model of our external environment stored in our memory, and that does not change. According to cognitivism, we never actually perceived the real world. Instead, we perceived an internal representation (mental model) of the real world (this view is sometimes termed *indirect realism* for this reason).

The problem here is, again, that such explanations are homuncular because, who looks at this inner picture? By definition, someone has to look at the picture to interpret it, update it, and act on it. Again, in the outside world this is not a problem. Who looks at a painting or photograph? *We* do. But, for the inner picture we need

---

* We return to this in the final chapter, but of course by analogy, the movements of the planets and the like can be described as if they are rule (law) bound. But, what are these laws? One can argue that they really exist in some metaphysical ether. Or, one can assume that instrumentalist theories of science are correct (as we do) and argue that these laws are merely constructs that human beings use for prediction. In physics and the like, it does not really matter whether one chooses instrumentalism or realism; the fact is that the law approach works. We can send spacecraft to the moon and the planets because we can predict their movements with a reasonable degree of accuracy. But, with psychology the situation is very different: very few (if any) human actions can be predicted with the lawlike deterministic accuracy available in astronomy. So, the implication that human behavior is *necessarily* law/rule bound becomes highly problematic.

a homunculus to look at and interpret the internal representation, who therefore needs another homunculus inside its head, who needs another homunculus, and so forth. So, mental models are conceptually impossible for the same reason that cognitivist rule-bound cognition is impossible.

It should also be noted that there are two types of mental model. The type normally proposed in safety/human factors is the pictorial mental model discussed in this section (this theory is most normally associated with Gentner and Stevens [8]). It should be noted that there is also a theory, proposed by Philip Johnson-Laird, that cognition itself is fundamentally the product of mental models [9]. However, these fail for precisely the same reason as the pictorial mental models: that is, who, precisely, operates these mental models? And, *what with* (in the absence of rules)? For further criticisms, see [10].

5.2.3 Experimental Data

Discussions of the coherence of (the pictorial) mental model stayed at the level of the abstract plausibility of the concept until recently, when experiments on change blindness closed the debate. To explain why, we have to remember that, according to cognitivism, we possess internal mental models of reality in our heads that are updated whenever certain mental alarms are triggered; that is, our brains are conceptualized as programmed such that if there is movement, then we pay attention to it, and then update our mental model accordingly. If we imagine a computer program watching something through a video camera, then it is clear that when certain criteria are met, it must react accordingly. So, in the case of a computerized burglar alarm, if the computer detects movement, then it triggers the alarm. Therefore, for a machine, to look is to see. If the information meets the programmed criteria, then the mental model is updated, and the response is triggered.

For the mental model theory to be true, therefore, to look and to see must be the same thing. This *is* the case for digital computers but not, it seems, for humans. For example, in one experiment subjects were asked to view a basketball-like game and follow the number of times one of the teams possessed the ball. Meanwhile, a woman in a gorilla suit walked briefly into the game. When asked afterward, subjects *did not recall seeing anything unusual*. Because it was not task relevant, the woman was not perceived [11].

An even more striking experiment was carried out by Fischer et al., who had professional pilots land an aircraft in a flight simulator. On various occasions during the landing approach, the simulator was programmed such that a large airplane appeared directly in front of them on the runway. Although the plane was clearly visible, two of the eight experienced pilots simply did not see it even though they were looking (at various points) straight at it (presumably because they were not expecting to see it) and simply landed their own aircraft on top of the jet. On later confrontation with a video of what happened, they were incredulous [12].

These are all examples of change blindness (so called because it shows that observers can be blind even to radical changes in the visual environment, even when they are looking directly at the object). Therefore, human beings can look and not see. In fact, if we are honest, then we have all occasionally been lost in thought and looked straight at someone without recognizing him or her. The story goes that a famous French philosopher, lost in thought, once walked straight into a lamppost. "I *saw* it, but I did not *realize* it," he explained after. Of course, we understand what this means, but for cognitivism to see and to realize were the same thing.

In opposition to the cognitivist view, the psychologist Kevin O'Regan and the philosopher Alva Noë argued the following: "Seeing is a way of acting. It is a particular way of exploring the environment. Activity in internal representations does not generate the experience of seeing. The outside world serves as its own, external, representation." In other words, we do not need a mental model of the world because we already have one in front of us. It is called the world. No homunculus is necessary because *we are the homunculi* [13].

5.2.4 NEUROPHYSIOLOGICAL DATA

Finally, there are more recent neurophysiological data that have again proved devastating to cognitivist orthodoxy. Boris Kotchoubey [14] carried out many experimental studies of event-related brain potentials, normally considered to be evidence for internal cognitive processing. These can be measured neurophysiologically.

To understand the significance of his experiments, we have to look at another corollary of the cognitivist view of perception and compare it with what one might term the sensorimotor view proposed by O'Regan and Noë (and Kotchoubey). For the cognitivists, there was an internal mental model that was continually updated via information passed to it from the outside world. These data were then stored and processed (via rules), and when the incoming data were perceived relevant, the organism reacted. In other words, first we perceived things (i.e., the information comes in and is operated on via rules in a mental model), and then we acted. This was, therefore, a two-stage model of perception.

In the alternative approach that one of its pioneers (J. J. Gibson) termed *direct realism*, we actually perceive the outside world directly, and we manipulate it, move in it, and act in it [15]. In this view, perception and action are two sides of the same coin; there are not two stages of perception but one stage.

Luckily, both these views lead to differing predictions that can be tested experimentally. So, for example, numerous experiments have been carried out measuring event-related potentials, especially the P3 wave, which is associated particularly with higher-level cognitive function. Naturally, according to the cognitivists what one would expect is that if a (sufficiently interesting) visual stimulus was produced, first one would get a pause (for cognition) associated with the P3 spike, and then one would get an action. However, experiments have stubbornly failed to produce this result, showing instead P3 waves coinciding with the required action or even coming *after* the required action.

Other experiments have been carried out that showed the best predictor of reaction times is not the complexity of the decision required (as one might expect from cognitivist theories and, of course, digital computers) but instead simply the raw amount of information made available. When this reaches a certain level, people simply act, regardless of the complexity of the decision. In other words, it seems that when enough information is made available such that people *can* act, they *do* act; moreover, the P3 wave amplitude remained low, suggesting again that people were not cognitively updating their mental model but instead reacted directly to the information in the external world environment. Finally, and most devastatingly, visual experiments have been carried out in which subjects were given a target to look for on a screen. Other dummy visual phenomena also appeared on the screen at various times. The cognitivist theory obviously *must* predict that the most processing will be associated with seeing the actual target. In other words, the participant is perpetually updating the mental model, and when there is a match between the desired information and the perceived information, presumably there will be the internal equivalent of bells and whistles going off as our internal computer registers a match, which then provokes an action. However, this was not the case. Instead, the cortical activity associated with *behavior* was directly stimulated, but the higher-level features (i.e., those associated with cognition) dropped off (because this now indicated that the task is over, so there was no longer anything to prepare for). Again, this indicated that when information enters the brain, it stimulates action but not cognition (as that phrase is generally understood). In other words, we react *directly* to the information, not to an internal representation of the information [14].

There are actually many other problems with cognitivism, such as the symbol grounding and frame problem [16,17]. However, it is sufficient to say that, due to these and many other problems, we view cognitivism as irredeemably flawed. We would also argue that it is due to the fact that it makes no sense that cognitivism has not proved illuminating in either safety practice or theory [18].

5.3 INFORMATION PROCESSING

Like cognitivism, *information processing* is a phrase that has two meanings, a general one and a specific one. Unfortunately again, the two are not often sharply differentiated. In its broad use, information processing can be simply another word for cognitivism. Even more broadly, it can be used to refer to *any* theory of mind that sees cognition as basically computational (therefore, connectionism [discussed below] is not a variety of cognitivism because it does not see cognition as rule bound, but it *is* an information-processing theory because it sees cognition as implementable on a computer). In other words, views of cognition analogous to the representational view discussed above are information-processing views in this broad sense because they see cognition as a matter of information input to the brain and processed, and then information is output in the form of speech or behavior, a process that is conceptualized as linear and sequential (step by step) (i.e., input-processing-output, *not* all three at the same time).

However, in its tighter and more specific form, information-processing theories are theories of *memory*. In this interpretation of the phrase, there are three main information-processing models.

5.3.1 Information-Processing Views

The first of these models is the Atkinson and Shiffrin model [19]. This is sometimes known as the modal model and posits memory as consisting of two main internal stores, a short-term memory (STM) store and a long-term memory (LTM) store. Initially, data are input into the STM before passing onto the LTM. It is generally accepted that the STM has a capacity of roughly seven bits of information [20]. The LTM, on the other hand, has an infinite capacity.

However, there are a number of empirical problems with this view of memory, apart from the obvious one that positing a store with an infinite capacity is meaningless. For example, it has been shown empirically that it is perfectly possible for some brain-damaged patients to have a perfectly normal LTM but no STM [21]. Of course, this is impossible in the Atkinson and Shiffrin model.

Moreover, the most basic assumption of the STM concept (that the longer something is kept in the STM store the more likely it is to be moved into the LTM) has been shown to be not necessarily the case [22]. And, there are many other problems with the concept.

Of course, there *are* other experiments that, it is claimed, do show support for the modal model. However, these can usually be explained in other ways, ways that fit more closely with ecological studies of the way human beings function in their natural environments. For example, experiments have shown that it is easy for subjects, by using their fingers as *aides memoirs*, to improve the capacity of the STM immensely [23]. As Glenberg wrote (in an important article to which we will return), "This evidence might be interpreted as evidence for a new finger control *module*, but it seems more sensible to view it as a newly acquired *skill*" [24; italics added]. He continued: "[in recent years] much of the evidential basis for a separate short-term [memory] store has been eroded." Given that this is the case, it seems that merely renaming the STM working memory will not solve the problems (and working memory, because it is a homuncular theory, has problems of its own; see below).

It was partly as a result of such problems that the alternative, Craik and Lockhart, information-processing model of memory was proposed [25]. This model at least blurs and (under some interpretations) abolishes the STM/LTM distinction and posits, instead of the digital (binary, quantitative) modal model, an analog, qualitative model in which memories are processed at different levels. Symbols that are processed at a deeper level are more likely to be remembered than symbols processed at a higher level.

However, despite the fact that the Craik and Lockhart model, in this respect at least, is more likely to be an accurate model of memory than the modal model, it has problems of its own, not least the extent to which it still posits memory as a matter of the storage of symbols. In this respect at least, both of these information-processing models owe something to cognitivist models of cognition.

But, because of this, they are open to the same objection as raised to cognitivist models (above): they are *homuncular*. Who precisely, uses the rules to obtain memory objects from the memory stores? Who checks that they are the correct objects? Who then understands that they are the right objects and passes them on to consciousness? (In fact, not only are these theories homuncular, but also they have a whole new set of problems that are even more intractable; see [26]. However, this follows from the homuncular nature of the hypothesis.)

5.3.2 EMPIRICAL DATA AND INFORMATION PROCESSING

Thus ends the discussion of philosophy. What do the empirical facts say about the memory store ideas? Bekerian and Baddeley conducted a classic study of how new radio frequencies for radio stations were remembered. Despite intensive bombardment by the media regarding the new frequencies, people did not on the whole remember the information; this suggests, of course, that the idea of a passive memory store is incorrect [27]. Elizabeth Loftus [28] and coworkers also carried out many experiments that showed the plasticity of memory, such as that suggestion can lead to the memory that the subject got lost at the age of three [29] or that the subject was born left handed [30].

However, the most devastating argument against the memory store/memory trace idea was an experiment carried out by Nader in 2003 on rats and other animals. The rats were electroshocked before putting them into a box, and in a standard stimulus–response way, the rats soon began to associate the box with shocks and to express fear before being put into it. In other words, they remembered that they were shocked every time they were put in the box. Then, the rats were given a drug that prevented creation of new memories (i.e., after they had created the association of box and electric shock). As one might expect, this made no difference; when showed the box, the rats still showed fear. They still remembered the association of box and electric shock. The memory trace, it seemed, was stored, so why should the inability to create new memories make any difference?

However, if the rats were then shocked one more time *after* receiving this drug that prevented the creation of new memories, then on being shown the box, *they showed no fear* (this occurred even up to 45 days after the initial shocks were administered). In other words, what must have happened is this: after the ingestion of the drug and the shock-box experience, the rats could not form a memory trace of this last experience because this was prevented by the drug. But, as well as this, the drug also erased the *previously stored* (box-shock) memory as well. How could this be? It could only be the case if the processes involved in creating a memory and retrieving it from storage are in fact the same. In other words, the memories of rats are not passive acts of retrieval but active acts of imagination. When we remember something, we are also actively *creating* that memory. Does this process also occur in humans? The answer seems to be that it does [31]. Memory is an active act of the imagination.

Finally, Kotchoubey related studies of brain waves of human participants carrying out simple memory tests, in which they were presented with a small set of objects and then, after a short period of time, were shown various objects and asked whether they had been in the original set. (So, there might be a set of cars or colors; after two minutes, participants were shown some of the same cars or colors [as well as some that were different] and asked if they had been in the originally presented set.) These experiments did *not* give evidence for internal processing (i.e., search) of memory stores but instead of immediate (under 300 ms) recognition of familiar stimuli, with complex cognition only taking place when subjects were not sure whether they recognized the object. These data are not compatible with cognitive models of symbolic storage but *are* assimilable to a noncognitivist model (in this case, Gibsonian ecological psychology) [32].

5.3.3 The Relationship between Information Processing and Cognitivism

Information processing and cognitivism are not quite the same theory, but it is clear why they need each other. Cognitivism needs the concept of a memory store to make sense. For cognitivists, when not used the rules and symbols have to be stored in a memory store before retrieval and activation. It has to be a store because the rules have to be brought out of the store in the same state they were put in; otherwise, they would quickly become useless. The same goes for symbols. If we use a mental model to cognize, it has to be the same mental model each time, so it has to be stored and retrieved looking more or less the same each time. If memory is not a store, therefore (and it is not), then the rules would quickly become useless as we would not retrieve the same rules from the memory store as we put in. Likewise, the information-processing view of memory needs to postulate the existence of rules to operate, as it were, the stores. But, if cognition is not rule bound, then that whole theory also breaks down. So, if one view falls, the other also falls.

5.3.4 Modularity

Another of the key assumptions of the cognitivist view was that cognition was modular; that is that the mind/brain could be meaningfully decomposed into semi-autonomous modules (this view was argued forcefully in Jerry Fodor's book, *The Modularity of Mind*) [33]. However, with reference to the Glenberg article, "What Memory Is For" [24], we have alluded to a reconceptualization of the evidence for the existence of modules, such that specific abilities could be seen not as the expression of *modules*, but instead as *skills*. The difference is fundamental. Modules are *internal* structures (so, for example, an STM store containing symbols would be an example of the modularity of memory). Skills, on the other hand, are orientated outward: one masters a skill *in the world*. Moreover, conceptualizing cognitive phenomena as skills highlights the extent to which they vary over time and the extent to which they can be learned. Modules are binary or digital; you either have them, or you do not. But, skills are analog; one possesses a skill (e.g., to play football or to ride a bike) to a certain extent; it is not an all-or-nothing (binary) phenomena.*

For example, in a number of experiments, the extent to which *attention* is a skill was demonstrated. Spelke et al. [34] demonstrated that, with practice, subjects could learn to read (and understand) short stories while at the same time taking dictation from spoken language. This not only demonstrated that attention was not a single-channel phenomenon but also indicated that

---

* In other words, it is fuzzy in the sense we discussed earlier.

attention … is based on developing and situation-specific *skills*. Particular examples of attentive performance should not be taken to reflect universal and unchanging capacities. … Indeed people's abilities to develop skills in specialized situations is so great that it may never be possible to define general limits on cognitive capacity. [35; italics added]

This was backed up by further experiments by these authors, who provided additional evidence that automaticity (requiring no cognitive capacity) or alternation (extremely rapid channel switching) was not involved. In other words, attention was a skill that could be learned, improved with practice, and had, in theory, unlimited capability.

This theory was explosive, not only because it challenged the dogma of internal modules for cognition with limited capacity, but also because it challenged the view of the mind as a digital computer. Instead, Spelke et al. saw memory as a kind of skill, and they compared it to the skills acquired by athletes. In other words, the brain is not best compared to a computer; instead, it is best conceptualized as a *muscle*. In the same way one develops muscles via practice in the gym, the more one practices something, the better one gets at it [36].

There are two reasons why this matters. The first reason is that it leads to fundamentally different predictions regarding the behavior of operators than cognitivist models. For example, Gopher in 1993 found that trainee pilots who practiced their attention-dividing skills using a computer game were twice as likely as a control group to become qualified pilots within 18 months [37]. In other words, given that attention is a skill, with practice they learned to master this skill, which led to greater ability to fly a plane. Viewing cognition as a skill rather than as a module leads to specific intervention strategies for training and action, intervention strategies that would *not* be considered if the view of cognition as modular was accepted.

The second and more subtle point is the light that this throws on the nature of expertise. There are two senses in which this is true. We deal with the first sense in this chapter and the second sense in Chapter 8. To deal with the first sense, what this implies is that to do something *a lot* means that (with motivation) one will become good at it. Here, the analogies should perhaps not be taken from mastery of higher-level cognitive tasks but from sports psychology. The evidence here is overwhelming that to do something a lot is to become good at it, and that the more you practice, the better you tend to become (with, to repeat, adequate motivation) [38]. What this suggests, to push this point slightly further, is that there is a certain corporeal aspect to task performance that cannot, perhaps, be quite caught with abstract models. If one compares driving a train or a car or even being an operator in a nuclear power plant not to abstract skill acquisition, but instead to, say, playing football or basketball, then different implications for training and learning present themselves. Moreover, it then becomes problematic to what extent one could actually learn these things *without actually doing them*. Does anybody think that one could learn to play football or tennis by being taught the rules in a classroom? Surely, only practical activity (developing the brain in precisely the same way one would develop muscles and fitness) can lead to real mastery of these activities? And, if this is the case, then this has real implications for training for driving a train or car or becoming a nuclear power plant operator.

The idea that this might not be the case, of course, descends from the view of TR detailed in Chapter 1, in which, reproduced throughout the entire Western education system, abstract knowledge is privileged over physical knowhow. Having raised this point, we now leave it again, to be discussed later.

There is one more point to be made. If you play football a lot, then you will get to be a good football player. If you play basketball a lot, then you will get to be good at basketball. But, apart from general issues of fitness, there is no particular reason to think that basketball players will be good at football (still less at judo, archery, or water polo) in the same way as there is no particular reason to think that someone who plays the drums will necessarily be any good at playing the saxophone (note that of course they might become good at all these things with practice, but what is discussed here is precisely to do something *without* practice, to assume that skills learned in one situation are easily transferable to another). To practice something is to become good at it, and if you do not do it, you do not become good at it. Skill is to learn to perform an activity, and activities are specific to the game (or situation) in which they are played. This is something else we return to later.

Until recently, the cognitivist or (nonconnectionist) information-processing views were the only ones available in psychology. However, alternatives have arisen that fall under the rubric of the new psychologies. These seemingly disparate approaches (situated, distributed, and embodied cognition; discursive psychology) in fact have large elements in common. All, we argue, provide a more useful basis for safety management than the more traditional view.

5.3.5 CONNECTIONISM

It should be noted before we continue that there *is* an information-processing model that does *not* fall victim to the objections above: connectionism. Unlike cognitivism, connectionism does not posit the existence of internal rules or a homunculus. Instead, it proposes cognition as the activation of interacting neurons; the view of memory it suggests is also compatible with the experimental data discussed above.

We have discussed connectionism elsewhere, and there is no need to discuss it in detail here [39]. However, it is sufficient to say that, because connectionism does not posit cognition as rule bound, it does not fall prey to the homunculus problem. Be that as it may, because connectionism still views cognition as computation, there are certain aspects of cognition that it ignores. These are now discussed.

5.4 SITUATED COGNITION

Situated cognition has its origins in the writings of the philosopher Martin Heidegger (1882–1969) [40]. Heidegger argued that, contrary to the suppositions of Cartesian (TR) psychology (which, remember, posited a radical disjunction between the person [subject] and the external world [object]), human beings existed instead in a "being-in-the-world." In other words, instead of a disembodied, abstract person who floated in some kind of metaphysical ether, human beings were always *in a context* or *situated* (a similar concept is used in Gibsonian ecological psychology;

the idea that there is no such thing as an animal, there is only ever an "animal-plus-an-environment").

For example, a train driver, instead of an abstract individual who happens to find himself or herself driving a train, is instead viewed as a driver-in-a-train. A pilot is a pilot-in-a-cockpit and so on. To use the language of systems theory (see Chapter 6) for a moment, what we are faced with is not a pilot who happens to be in a cockpit, but instead a pilot-cockpit *system*. Therefore, when looking at the pilot in the cockpit, the unit of analysis is not two things but one thing: the system. There may well be other people in the cockpit. Again, the same principle applies. A pilot and a copilot in the cockpit are not best seen as three things but one thing: the pilot-copilot-cockpit system.

To see how this differs from the cognitivist view and, importantly, how this has an impact on the creation of taxonomies, we should look briefly at what is probably the most widely used taxonomy in the field of safety science: Rasmussen's skill-rule-knowledge (SRK) taxonomy. We discussed this taxonomy elsewhere, and there is no need to repeat that discussion here [41]. However, it is enough to say that the assumptions of SRK are broadly cognitivist; it posits internal rules and mental models as the key engines of cognition.

In other words, SRK assumes that there is a radical disjunction between the internal and the external world. The external world is outside and material. The internal one is like a digital computer and is made of mind stuff (the fact that it *ultimately* may well be supposed to be material is irrelevant on this level; for all intents and purposes, the outside world consists of matter, the inside world of information). The result of this is that your situation does not really matter, internal cognitive processes cause your behavior. Whether you are in a cab, in a cockpit, or sitting at home having a beer, the rules and mental models will pretty much work in the same way. Your cognition and behavior, according to SRK, is not situated.

This is such a deeply rooted idea in Western culture that many people have great difficulty in seeing that it is highly suspect. Seeing cognition as unsituated is like attributing an internal locus of control. In other words, it is internal, private cognitive processes that "cause" your behavior. The situation you find yourself in is less important.

However, this view leads to inaccurate predictions. In a classic experiment (usually referred to as the Good Samaritan experiment for reasons that will become obvious), Darley and Batson [42] studied the behavior of seminarians at the Princeton Theological Seminary. First, they gave all participants a questionnaire that assessed the reasons for their religious vocation (in which they were given the option "to help my fellow man"). They then asked the subjects to prepare a short talk on preset themes (including that of the Good Samaritan) and then to walk to a nearby building to present it. On the way, they passed a "stooge" who was paid to act as if he had been mugged or was otherwise overcome with illness.

The experiment was in two stages. First, randomly selected participants were told about the experiment and were asked, who would help the injured person? The answer most people in the West give is that the participants who stated that they wished to help their fellow man or the participant who was selected to give the talk on the Good Samaritan would help him. This is the standard (nonsituated) view: internal (cognitive) phenomena cause behavior. If you are a good person (being good posited as being a stable, internal aspect of your personality), then you will help people.

Unfortunately, this view does not hold up to scrutiny. The *only* worthwhile predictor turned out to be whether the participant was in a rush. Many of those who had stated that their only wish in life was to help their fellow humans actually stepped over the stooge (who, remember, may have been dying for all they knew) to hurry on their way [42].

In other words, the behavior of the seminarians was *situational.* Seminarians-in-a-rush will behave differently from seminarians-not-in-a-rush. Internal (cognitive) factors seem to be less important. This finding has been backed up by many famous studies and experiments, with the most notorious example the well-known Stanford Prison experiments. In these studies, Philip Zimbardo set up a fake prison in the university's psychology department. Then, 21 volunteers (healthy and normal according to a battery of psychological tests) were divided *randomly* into prisoners and guards. The guards almost immediately started to behave like hard-bitten disciplinarians, becoming increasingly sadistic and brutal as the experiment progressed, even despite the fact that some of the guards had previously identified themselves as pacifists. Again, we find the *situation* (not internal dispositions) as the key predictor of behavior [43].

However, that last sentence is in some ways slightly misleading. We would not want to start playing the stimulus–response game and seeing the external environment as causing behavior as if we merely had to change the situation to change the behavior. Instead, we should see, as always, a situation-person *system*, which has emergent properties *of the system.*

This can be made clearer by another classic study. Hartshorne and May carried out studies on children's propensity to cheat. Young children were given simple written tests (of a variety of types) in a variety of different situations such that cheating was either difficult or easy. Propensity to cheat was situation specific (not only in relation to the physical situation, but also in relation to the type of test; children would cheat on some tests but not others, *ceteris paribus*). Although it *was* true that while there were sociocultural biases (social class, age, etc.) that functioned (to a certain extent) as predictors, these were only really predictive when triggered by the situation. Again, cheating behavior is an emergent property of the child-test system [44].

These studies are vital for an understanding of safety behavior. A good example of the implication of studies like these is that they provide strong evidence for the view that human error is *also* situational.

5.4.1 "Human Error" and Situated Cognition

The opposing thesis to the situational explanation of mistakes or errors is that there is an internal dispositional tendency known as *accident proneness* that causes accidents. This theory is based on the statistical fact that, in a large organization, some people will be more associated with accidents than others will. Rather than seeing this as a feature of the statistical distribution of accidents in a system, some writers concluded that some people have an inherent tendency to be accident prone.

The problem with this individualistic view is that even when employers replace repeat offenders, accident rates tend not to go down [45]. Moreover, attempts to discover what these internal dispositions are and how they might be used to predict propensity toward accidents have not produced meaningful or replicable results [46].

Once more, the *situation* is the most effective predictor of accidents. Individual differences do exist, but they tend to balance out over time. In work that is dangerous and hard to learn, people will tend to have more accidents than when the work is easy to learn. This is not to say that the human is irrelevant. People are different and have different reactions to specific situations. Moreover, people learn, and some people learn quicker than others do. People who learn (some tasks) quickly will tend to be better (and therefore safer) at these tasks more quickly. (However, it should be remembered that this is a transient situation. That is, by the time slower learners have caught up, the quick learners may well then be bored, which leads to different error types.) In the same way, errors caused by stress, fatigue, inattention, and so on will tend to be spread throughout the workforce primarily through systems issues. It is only if the *whole workforce* is overstressed or overworked (in other words, if it is a systems issue) that prediction follows. But, here the issue has become a collective (systems) one rather than one of an individual cause–effect. Again, errors are better seen as emergent properties of the person-work-environment system.

5.4.2 So What?

So, actions are situated in a specific social situation. But, so what? What difference does this make? To see what difference it makes, we can turn to one of the pioneers of situated cognition, Lucy Suchman, and her classic analysis of what might (in another context) be called human error. Here, we consider two people, A and B, attempting to photocopy a document. Specifically, they are attempting to make a two-sided photocopy of a bound document.

Remember, those in the cognitivist tradition have assumed that human beings are sequential rule followers. For example, carrying out a driving task means following rule one (open car door), then proceeding to rule two (sit down in car seat), rule three (fasten seat belt), and so on. Importantly, designers and technicians in this tradition have tended to build machines predicated on this idea. The problem, as we have seen, is that human beings are *not* in fact sequential rule followers.

People do use rules, of course; these are rules such as drive on the right- (or left-) hand side of the road, remember to close the fridge door, don't press the big red button except in an emergency, and so on. However, these are not the internal algorithmic rules posited by cognitivism. On the contrary, they are socially structured rules learned in a social situation. As the philosopher Ludwig Wittgenstein noted, when we interpret rules in the same way as other people, this is because we have certain cultural assumptions that we all share (what he called forms of life) [47]. Ignoring this can lead to confusion.

In Suchman's example, the users need to use the photocopying device known as the bound document aid (BDA) [48]. To begin, they read the instructions: "To access the BDA, pull the latch labeled Bound Document Aid." Person A finds it and points to it, saying "Right there." Person B then puts his or her hands on the latch and pulls. However, nothing happens. Person A repeats the instructions, then suddenly sees an illustrative diagram and adds, "Oh, the whole thing!" Person B then succeeds.

The original human error (Person B failing to operate the latch properly) is not an individual error at all. It emerges from a situation in which we have two people and an ambiguous rule. Person A did not realize that when the rule stated "lift the latch" this actually meant the whole latch and not just part of it. It was only when Person A happened to see the diagram that the intention of the designer was revealed. It simply had not occurred to the designer that the rule was ambiguous (i.e., indexical, a word that will be explained later in this chapter). Things like this are often classified as human error from outside (the system), but in fact viewed *from within* there is nothing inherently erroneous in the behavior.

Finally, A and B manage to locate the correct place to put the document and then press "start." However, there is now a new problem: the machine copies the document to the wrong size of paper (of course, from the point of view of the machine, it is the right size of paper, but A and B have not realized that it has been preset). They set the reduction feature, and press "start" again, and successfully copy an *un*bound master copy. However, they now have to bind this and make it double sided. They go through the BDA procedure again and bind the document (which is not yet double sided) and find that the machine has somehow misprinted such that the pages are now not in order. So, they try again.

Now, it is important here to consider things from the point of view of the humans and the machine. From the point of view of the humans, they are still trying to make a bound, two-sided copy of a document. However, of course, the machine does not know about the mixed-up page numbers, and its internal sequence tells it that the whole process is now over, and that therefore A and B appear to be trying to start the whole process again: copying from an *un*bound document. Of course, the machine will now not accept the bound document, and so nothing happens until eventually (and inevitably) A and B start pressing buttons more or less at random to make the machine work. Eventually, they have to give in and start the whole process again [48].

This failure — a classic human error — turns out to be a mismatch between the idea the machine designer had of how humans are supposed to think and how they in fact do think. The designer assumed that because the machine proceeds along algorithmic or cognitivist lines that therefore the humans would as well. Needless to say, a cognitive ergonomist who starts from the assumption that there is human error involved in a failure situation will tend to partition out the behavior of the people and the static original design of the machine. The human behavior tends to be explained by reference to individual psychology in terms of mental models or cognitive maps. The design is viewed as separate (maintaining the subject-object distinction) and as a technical aspect (the mental model of the expert designer is rarely alluded to).

There is another individualistic approach to this human error (i.e., one that plays down interaction between person and environment): to focus on the unfamiliarity of A and B with the task. But, this fundamentally misunderstands what rejection of cognitivism implies. If one rejects the idea of internal cognitive modules in favor of skills that one learns, then it is clear that the brain's key function is to learn, and it is our capacity to learn that differentiates us from the animals (not the fact that we can learn, but the extent to which we can learn). Looked at from this point of view, we never stop learning.

This is particularly true of the workplace. If we stop seeing work activities as separate modular activities but instead as part of a dynamic open system that is composed of our workmates, the changing socioeconomic system, and the external environment, then it becomes clear that we never could *fully* learn any job because there will always be new phenomena that have to be experienced and handled. Again, learning is not a binary or digital phenomenon but an analog one. We learn things to a certain extent. We never simply know or don't know how to carry out a task. The corollary of this is that we never truly master a task (most genuine experts agree, proclaiming that, for example, you never master playing the guitar or painting; the more you learn at these tasks, the more you learn about what you do not know).

Given that this is the case, this means that every example of human error can in an absolute sense be seen as a learning problem (in the sense that it is an activity taking place in a learning context). This again calls into question the meaning of the whole concept [49].

5.5 EMBODIED COGNITION

Embodied cognition is the second strand of the new psychologies we deal with here; again, it has its roots in the thought of Heideggerean phenomenology (especially as it was developed by Maurice Merleau-Ponty).* Despite the different name, embodied cognition is really very similar to situated cognition, such that some people have argued that they are in fact the same theory. That is possibly true, but they have different emphases. Both emphasize their opposition to Cartesian dualism and the sharp distinction between the subject and the object. However, embodied cognition is much more concerned with the extent to which the division (i.e., taxonomic distinction) between the mind and the body is a false one. Again, the concept of emergence is key here: embodied cognition views the mind as an emergent property of the brain-body *system*. In this view, there is no sharp distinction between our bodies and our minds. We literally think with our bodies (in other words, as fuzzy logic tells us, the distinction between mind and body is analog, not digital).**

Again, we come to the "so what" aspect. The point is that cognitivist taxonomies that neglect this feature are only ever going to come to a partial description of human

---

* However, more recently embodied cognition is more associated with the linguist George Lakoff, the philosopher Mark Johnson, and the biologists H. R. Maturana and F. J. Varela.

** In the same way, it would be a mistake to see us as arguing for a subjective approach to science as opposed to an objective approach. Instead, we see subjective and objective as a fundamentally analog distinction on which we sometimes impose a digital structure for pragmatic reasons. Our interior world is socially structured, but there is also a dynamic interpretation and creation of our view of the external world.

behavior. Some very simple examples taken from standard ergonomics textbooks can illustrate what we mean.

5.5.1 CONTROL

The salient point in all these examples is the concept of control. Human beings attempt to self-regulate and achieve what is known (in physics, in a different context) as the Goldilocks' point. This may be the point between not too hot and not too cold or between not too bored and not too stressed, not too hungry and not too full, and so on. They manipulate their environment in an attempt to achieve this. The implications of this are pursued in the next chapter.

5.5.2 FATIGUE

There are few more experimentally validated results than the fact that fatigue can have a deleterious effect on performance. For example, Grandjean found that, among telephone operators, performance in various tasks deteriorated after work when compared with before work. Similarly, Broadbent carried out classic experiments on attention in which he discovered that performance at a simple attention task (noticing a specific signal) was greatly reduced by fatigue [50].

5.5.3 BOREDOM

An aspect that is closely associated with fatigue is boredom. Again, numerous experiments on drivers, pilots, and train drivers have shown that boredom can lead to lowered performance, and that being bored and tired (fatigued) can lead to an emergent problem in which performance is radically reduced. The classic example in this case is the example of the Australian road train, extremely long trucks that drive through the outback. Despite the fact that it is far harder to drive these vehicles in urban situations, accidents are rarely reported in an urban setting. Instead, by far the most common type of accident is when the driver simply drives off the side of the road in low-traffic situations. It seems here that a state of highway hypnosis has followed from a long period of low-stimulus driving leading to fatigue, boredom, or both [51].

5.5.4 STRESS

The corollary of this is performance reduction because of stress. Callaway and Dembo tested subjects' ability to judge the size of various objects while under the influence of amyl nitrate, which produces a high (or high arousal in the jargon). They found that these subjects tended to make more mistakes in terms of estimating the size of the objects than the control, and that these mistakes were associated with narrowed attention [52]. Baddeley also carried out numerous experiments on deep sea divers that demonstrated the negative effect of stress on performance [53].

Here, the basic point is obvious: the more stressed a subject is likely to be (allowing for many confounds), the less well they are likely to do. Calmness is associated (generally) with maximum performance, with boredom/fatigue associated with, again, performance decrements. But, it must be stressed (pardon the pun) that what we are talking about here is embodied states based on brain states and physiological responses in the cases of both boredom and stress, not malfunctions in unobservable cognitive architecture.

5.5.5 OTHER ISSUES

There are numerous rather obvious issues that relate to performance, including access to food, fluids, light, heat, and various other factors. It would be a mistake to consider these environmental issues as somehow extrinsic to real performance. Instead, in line with situated cognition, we should see this more as a systems relationship, with the operator attempting to self-regulate while in an environment that is too hot or not hot enough, where the operator is bored, stressed, or whatever.

Two major corollaries follow from this: again, taxonomies of human behavior and human error that omit embodied features, the real physical context in which the operator finds himself or herself, are omitting a great deal. Second, given that in a poor environment the operator has no choice (as it were) but to have his or her performance reduced (possibly leading to error), this gives us another attributional or causal pathway to follow, which leads us away from the influence of alleged cognitive structures in the head and, again, toward the environment.

5.5.6 MISTAKES AND THE HAMMER

There is one final point to make about embodied cognition. We mentioned that embodied cognition developed out of the philosophy of Martin Heidegger, among others. Specifically, Heidegger used the classic example of the hammer to develop his anticognitivist psychology [40]. Heidegger pointed out that we know a hammer is a hammer not because it corresponds to some image of a hammer we have in our heads, but because we use it for hammering. Actually, most times we do not even go this far: we simply pick up the hammer and start hammering, and we are not *aware* of it *at all* (i.e., we are not aware of it *as a hammer*). But, sometimes the hammer breaks, and *then* we suddenly become aware of the hammer and have to think of a way to solve this problem of the broken hammer, a situation Heidegger called breakdown. Breakdown brings the hammer to our attention, and in fixing it, we learn more about the hammer, how it works, and possibly how it can be made to work better. Breakdown, mistakes, and accidents can lead to knowledge and understanding.

Why is this important? The importance is that it challenges one of the key assumptions we have about error (human or otherwise): *it is always bad thing*. In an important article, Rauterberg quoted Napoleon, who once remarked: "I learned more from my defeats than my victories" [54]. Why might this be the case?

After noting (correctly, for reasons discussed earlier) that it is probably meaningless to talk about what percentage of accidents were caused by human error, Rauterberg goes on to show that the cognitivist rule following assumptions about human behavior are at odds with Napoleon's contention. Cognitivists tended to assume that the inner mirrors the outer, and that therefore the novice began with a *simple* mental model that then became progressively more complex as they mastered the task. But, Rauterberg presented experimental data that showed that the behavior of novices at a given task is *more* complex than that of experts. What this demonstrates is that people adopt a deliberately broader range of behavioral strategies than is strictly necessary when beginning a task (in other words, they deliberately try things out and experiment with the system or, to put it another way, deliberately make mistakes) to get to know the system. Then, they will make mistakes that will enable feedback to be gathered, which will help them to understand the system [54].

Yet again, we find cognitivism provides a misleading picture of human learning. If we accept the theory of mental models, then mental model theory makes incorrect predictions (as above). However, if we accept the other major alternative (mental logic or rule-based cognition), then we are faced with the problem that human beings *deliberately* break the rules, and the reason they do this is very simple: to find out what will happen if they do.

Under the cognitivist view, which presupposes that there is one way of performing a task, that this is rule bound, and that following the rules therefore defines rationality, this would of course be considered irrational. However, Frese et al. described experiments in which novices were divided into two groups in a computer task, a group that were encouraged to commit errors and another group prevented from making errors. The group prevented from making errors learned more slowly and demonstrated reduced competence compared to the other group. This study has been replicated in various contexts, including sport. Frese used the Heideggerean concept of breakdown (as seen above) to explain these findings. Making mistakes can be a creative act, one that can lead to greater understanding of the system. Given that, therefore, we continue to learn throughout our working life, it seems that committing errors is an *inevitable* and *desirable* aspect of the working process (as can be seen from the photocopying example above) [55].

Even in a task we know backward, we sometimes *deliberately* have to commit an error to maximize our performance. As Weinberg and Weinberg put it:

> The lesson is easiest to see in terms of ... an icy road. The driver is trying to keep the car from skidding. To know how much steering is required, she must have some inkling of the road's slickness. But if she succeeds in completely preventing skids she has no idea how slippery the road really is. *Good* drivers, *experienced* on icy roads, will intentionally test the steering from time to time by jiggling to cause a small amount of skidding. ... In return they receive information that will enable them to do a more reliable, though less perfect job. [56; italics added]

It is only good drivers who commit the errors. An inexperienced driver would not, and that, therefore, would be the error. This finding, it is safe to say, can be extended across a whole spectrum of work-related activities. For example, as Anderson put it,

> Consider an instance of (fairly) simple tool use: using a paper clip to retrieve a life-saver from between the car seats. What we *don't* do is exactly pre-compute the distance of the goal, the required shape the paper clip must take, and the precise trajectory of approach necessary to hook the life-saver. Instead we unbend the clip according to a rough estimation of the requirements of the task, try it out, bend it a bit differently. ... The cognitive strategies employed are constrained and shaped by the performance

characteristics of our body as a whole in the given circumstances. … The shape, size, and flexibility of our fingers first rules out the possibility of a tool-free approach … then the size and configuration of the surrounding space becomes a further restraint. … this in turn limits the range of possible view of the object. … Finally, having determined some provisionally acceptable definition of the problem space, we (re)shape the clip and go about our iterative work. Here cognition is bound up with the shape, size and motor possibilities of the body as a whole and further relies on the possibility of repeated interactions with the environment. [57]

Taxonomies and analyses that ignore this and posit cognition as a matter of internal information processing or have categories such as human error as if this exists in a vacuum will produce incorrect intervention strategies. In the case above, it is not that the subject lacks information in the abstract sense or that he or she suffers from cognitive limitations. Instead, the subject has to be allowed to use his or her body in an active, exploratory role to gain dynamic embodied information. It is clear that making mistakes will be part of this process, and that these mistakes will produce important feedback that will lead to the completion of the task.

5.5.7 TRAINING AND LEARNING

Another important conclusion from this adoption of the situated and embodied approach is a different approach to training and learning. Too often, especially in the more advanced fields of engineering, practices are taught in the old-fashioned cognitivist way, in which decontextualized rules and formulae are taught in a class-room by a teacher, while the students are to sit back and passively process these data.

However, this is an approach that has been largely abandoned in the current psychology of education, which now recognizes situatedness. This approach is summed in the classic paper "Situated Learning and the Culture of Learning" by Brown et al., in which they summarize the differences between the old cognitivist approach and the new situated approach [58].

They begin with a simple distinction between knowing how and knowing that. To know that something is the case is merely book learning: abstract knowledge (like my knowledge that the Earth revolves round the sun, for example). Knowing how, on the other hand, is very much the kind of learning that we would wish to encourage in the workplace (and outside). The key point to understanding knowing that is to understand that learning (like all cognition) is always *situated*. For example, one does not learn abstractly how to drive a train, fly a plane, or work a lathe; instead, one learns how to drive *this* train (or plane, etc.) at this time, in this place. Learning how to use a tool (and trains, planes, and automobiles are tools)

involves far more than can be accounted for in any set of explicit rules. The occasions and conditions for use arise directly out of the context of activities of each community that uses the tool, framed by the way members of that community see the world. The community and its viewpoint, quite as much as the tool itself, determine how a tool is used. … Because tools and the way they are used reflect the particular accumulated insights of communities, it is not possible to use a tool appropriately without under-standing the community or culture in which it is used. [58]

This use has to be understood *from the inside*. Outside experts, no matter how well meaning, can never grasp the truth of this cultural use. In other words, experts in accident investigation, cognitive ergonomics, and the like can create boxes-and-arrows models or error flowcharts all they wish, but they never gain true insight into what it feels like to be a pilot, driver, or nuclear power plant operator. "The culture and the use of a tool act together to determine the way practitioners see the world: and the way the world appears to them determines the culture's understanding of the world and of the tools" [59].

Certainly, the Confidential Incident Reporting and Analysis System (CIRAS) database (see Chapter 2) has examples of this: the drivers call it route knowledge, which is simply the experience gained from actually traveling on the route. It cannot be gained by merely looking at pictures or reading about the route in books. The type of experience that is actually gained first hand is *authentic*.

It should be noted that a key point of the situated view of learning is the concept of apprenticeship, in which an older and more experienced member of the culture takes a younger member "under his or her wing" in an official or unofficial capacity and shows the apprentice how things are really done. This may, of course, differ, perhaps greatly, from the official story of how things are done but is sometimes formalized in mentoring-type schemes.

It is taken for granted in much writing on this subject that unofficial working practices are by definition incorrect, wrong, or otherwise sloppy. Yet, by what criteria would this be decided? The only logical answer is the criteria of *management*. Managers make the rules, and managers decide who has broken them.

But, management can get it wrong. In the book *Paying for the Piper*, Woolfson et al. quote the example of the *Piper Alpha* oil disaster, in which there were two sets of divers: one set at the time of the disaster followed the correct rule book emergency drill; the other had developed their own procedures, in defiance of management. The rule followers perished, yet many of the violators survived [60]. This is a classic example of the collaborative learning developed in a social context by people actually at the man–machine interface. If it is granted that situated cognition is a valid representation of human learning, it is easy to see why collaborative learning has a good chance to be more valid than strategies thought out by people (experts) from *outside* the culture, who do *not* understand the specific problems encountered in this specific work environment. The implications of this are rather radical. It is often argued that there is a crisis of communication in many organizations. What we are suggesting here is that the crisis is as likely to be a problem of bottom-up communication (workforce to management) rather than, as is more usually assumed, a failure of the *workforce* to listen. Again, this is a point to which we return in the final chapter.

5.6 DISTRIBUTED COGNITION

Whereas situated and embodied cognition derive from the work of Martin Heidegger and others, distributed and discursive psychology (see Section 5.7) are descended from work by the philosopher Ludwig Wittgenstein, the psychologist Leo Vygotsky, and the sociological tradition generally (especially the school of ethnomethodology).

These schools of psychology emphasize the social aspects of cognition and argue that language and behavior are fundamentally social phenomena. This does not contradict embodied and situated cognition, but it does lead to a different emphasis of approach.

Distributed cognition has its roots in Vygotsky's school of activity theory psychology and ethnomethodology, but in its modern format it was developed by the psychologist Edwin Hutchins in the 1980s and 1990s. Hutchins has argued that, whereas for most conventional (cognitivist *or* behaviorist) psychology the individual is the basic unit of analysis, more productive and useful insights might follow from taking the social group as the basic unit of analysis [61]. In doing so, Hutchins literally turned cognitivist psychology inside out.

Cognitivists argued that there were rules and mental models inside the brain. As we have seen, this is not the case, but, argued Hutchins, it *is* true to say that there are social rules that exist outside us in the social world. These are the rules that govern our life; for example, there are rules about how to play games, about how to drive, or about how to carry out tasks in the workplace. These are not algorithms, stored as in a computer. But, they are real nonetheless, as anyone who wishes to try breaking these rules (e.g., by walking out of a shop with expensive goods without paying for them) will find out soon enough.

In a similar way, there are what one might term social mental models: cultural and social filters that constrain what we can and cannot see. Again, these are not programmed in the brain but are instead emergent aspects of social interaction.

Hutchins argued that this viewpoint can provide an insight into much that was traditionally classed as human (i.e., cognitive) error. To see why this might be the case, we wish to take a close look at his classic work with Tove Klausen: "Distributed Cognition in an Airline Cockpit" [61].

5.6.1 DISTRIBUTED COGNITION IN AN AIRLINE COCKPIT

The work of Klausen deals with a very detailed analysis of video footage taken in a flight simulator of an experienced flight crew taking off. It is significant here because a minor mistake or error takes place, and therefore we can see the extent to which distributed cognition might be useful in terms of analysis of this form of behavior.

To begin, here is Hutchin's transcript of what the flight crew said (S/O = second officer; Capt = captain; F/O = first officer):

```
0216     S/O        xxx nasa nine hundred
0224     S/O        departure report
         S/O        nasa nine hundred from eh sacramento to los angeles
                    international we have eh /.../ fuel on board twenty seven
                    point eight fuel boarded is not available out time is one six
                    four five up time is one six five five
0247     Capt       oakland center nasa nine hundred request higher
{F/O reaches to vicinity of altitude alert setting knob when ATC begins transmission}
0254     OAK24L     nasa nine hundred /.../ roger contact oakland center one
                    thirty two point eight
{F/O pulls his hand back from the altitude alert knob when ATC says "contact oakland
center" 2.5 seconds after the end of ATC transmission, F/O looks at Capt}
```

{Capt looks at F/O}
0300   F/O   thirty two eight
          Capt   thirty two eight?
          F/O   yeah
          Capt   ok
0303   S/O   that's correct, nasa nine hundred Capt \one three two eight ah,
                       nasa nine hundred
{Capt twists knob on radio console}
{F/O looks in direction of Captain}
0315   Capt   center nasa nine hundred twenty one point seven for two three
                       zero requesting higher
0323   {S/O turns towards front of cockpit}
0325   {F/O looks at Captain}
0325   OAK15H   nasa nine hundred /.../ oakland center climb and maintain flight
                       level three three zero and expedite your climb please
0327   {F/O reaches the altitude alert as ATC says "climb and maintain"}
0330   {When ATC says "expedite your climb" S/O turns to the performance tables on
                       the S/O work surface}
0331   F/O   ok
0333   Capt
{Capt leans toward and looks at F/O}
                       I didn't catch the last part
0336   F/O   expedite your climb
          Capt   ok

What is going on here? Hutchins argued as follows:

> This is the sort of description that a pilot would give [of the transcript above]. As the
> crew approached the altitude to which they were cleared, the Captain called Air Traffic
> Control and asked for a clearance to a higher altitude. The controller handed them off
> to a high altitude controller who gave them a clearance to their cruising altitude and
> instructed them to expedite the climb. The Second Officer increased the thrust and they
> continued their climb. [61, p. 21]

But, he went on to argue that further analysis explains many of the basic concepts
of distributed cognition and showed why it is such an illuminating mode of analysis.
First, there is the importance of planning, again not in terms of algorithms inside
the head, but in terms of socially appropriate goal-directed modes of behavior.

At the time of the transcript, the plane is cruising at 19,000 feet (remember, this
is a simulation). However, for the next section of flight, the plane must rise to 23,000
feet, for which it needs clearance from ground control.

Therefore, the captain has to speak, but, importantly, everyone in the crew has
access to the frequency on which he normally communicates. The information he
makes available is therefore public. Hutchins reminded us that it is impossible to
predict where the information will actually go. Other members of the crew may be
listening. On the other hand, they may not, and this is an inherently ambiguous
aspect of the situation.

Now, we move to the social rules of the captain-cockpit system. As Hutchins
pointed out, conversations between the captain and ground control are tightly structured and consist of a small number of expected statements and responses. The
captain therefore puts in a request to climb, expecting back a statement stating that
this is all right. To repeat, at this point, the pilot does not know who has heard this

request. However, at this point the F/O moves his hand toward the altitude alert setting knob. From this, we must assume that the F/O has heard the request and is expecting (like the pilot) the affirmative reply from the control.

This is an example of a *social* mental model, a model, in this instance, of expectancies created within the crew. Information moves around the system, and the individuals within the system react accordingly. However, here this expectation is violated. Instead, the pilot is instructed to contact another controller at Oakland. Now, the social rules that govern this kind of interaction state that when the captain receives contact from control, he is supposed to acknowledge. But, he does not do so. Instead, he pauses for 2.5 seconds, after which the captain looks at the F/O but says nothing. The F/O then says "32 8" to the captain, who replies, "32 8?"

This is a good example of what Hutchins called *intersubjective agreement*: the shared social meanings that give meaning to any specific utterance. The statement "32 8" is *indexical* (a word explained in more detail below). However, for now it is enough to state that this means that its statement is context specific, and the context is defined by the social rules and representations common to those in the cockpit. It is these intersubjective agreements that provide the context such that both the captain and the F/O know that this refers to the frequency of the other Oakland control. Once this has been established, the captain acknowledges the control (obeying the social rules again of this particular kind of discourse) and reads the frequency back to the controller.

After reading back this frequency, the captain tunes the communication radio to the desired frequency and contacts the other controller. He is now slightly pushed for time after this delay. Now, information has passed to and from the members of the crew. What has happened is that, building on the shared mental model of expectancies, the captain expected a certain response from a controller. He did not get it, but, partly because he believes that all the problems are now solved and partly because he is now under time pressure, he (and his crew) have an even greater expectancy that he will get the appropriate response this time.

However, he both does and does not get the expected response. He instead gets the response: "Oakland center, climb and maintain flight level 330 and expedite your climb please." Now, the crew has a shared expectancy that they will get the first part of the message but not the second. However, this piece of information is passed on to the F/O and S/O but *not* the captain, in line with our statement that one never knows who has received information unless it is specifically acknowledged.

The phrase "expedite your climb" has a very specific meaning and leads to actions, in this case, that maximum engine thrust is to be used. The S/O immediately begins to perform this action. The captain also has a job to do, which is to repeat back his clearance to control. This he does; however, he only reads back the first section of the message ("330 nasa 900"). Crucially, he is *not* challenged on the fact that he fails to read back the statement "expedite your climb."

Therefore, we now have a breakdown in the system. Crucially, this is not anyone's fault. We can never know that everyone will hear every aspect of every part of information transmitted to a social system. There might be any number of distractions, hearing difficulties, whatever. Therefore, we rely on constant feedback to ensure that everyone is on message, but sometimes these feedback mechanisms (which of course are in the form of social rules) fail. However, this has now led to the situation in which the captain has, as it were, drifted away from the social mental model of events shared by the S/O, F/O, and controller.

Immediately afterward, however, the captain turns to the F/O and states: "I didn't catch that last part." Again, this is an indexical statement (as, it will be argued, all statements are); its meaning varies according to context. In this context, it means "Could you tell me what the last part of the statement by the controller was?" Crucially, there is a fundamental epistemological problem here. The captain is unable to state exactly what the last part was, and so the F/O has to guess what piece of information the captain lacks (in other words, the captain is asking: "Tell me the information I don't know," to which the reply could be "What information is that?" which invites the question, "Well, if I knew that I wouldn't be asking would I?"). Luckily, the F/O guesses correctly, and with a shared social mental model again, the crew can complete the change in altitude.

5.6.2 IMPLICATIONS FOR SAFETY

This was a minor error, which in any case occurred in a simulator. However, it does indicate the advantages of looking at the captain, F/O, S/O, controller group as a system, with information passed to and fro within the system. A cognitivist approach would simply see this as a lapse of attention and look for the cognitive mechanisms that led to it. But, a distributed cognition approach demonstrates that it is simply not realistic to assume that all information is always completely passed throughout the system (we see this view again in the next chapter), and that there is no need to explain this in terms of error. Instead, it is better to see it as a problem with information transmission itself and (in this case) a problem with the feedback loops that have been set up within the system to ensure that information *is* adequately distributed. In this situation, the captain was aware that he did not have access anymore to the shared mental model shared by the rest of the crew and was able to communicate this. The intersubjective knowledge of the F/O (such that he knew that the phrase "I didn't catch that last part" in this context meant, "Could you tell me what I missed?") meant that the captain could be reintegrated into the system. There are two other main points to be made here.

5.6.3 METHODOLOGY

One point on which we beg to differ from Hutchins is in his discussion of methodology. He argued that the use of what (we would call) taxonomy theory is invalid because reliability has connotations of objectivity, which he claimed is impossible (he is right on this point). However, we would argue that taxonomy theory and practice do not promise objectivity but instead offer the intersubjective agreement that he has demonstrated lies at the root of all social practice. Testing taxonomic reliability is simply measuring intersubjective agreement. There is no problem with developing a taxonomy and using the data produced to (for example) compare these data with other data and produce a statistical analysis. Analysis of video data by coders is still an unexplored field, and we would argue that the methodology of taxonomy theory and the implications of distributed cognition coincide nicely.

The second point leads us to our final section on indexicality. Cognitivism tends to downplay context, as we have seen, and so even simple discourse as above seems complex, perhaps incomprehensible with a cognitivist approach. In the above example, the results were trivial, and in any case, the crew was in a simulator. In other situations, the results can be more serious. For example, in 1977, there were 583 people who died when two Boeing 747s collided on a runway in Tenerife in the Canary Islands. Immediately before the crash, the pilot had radioed to the controller: "We are now at takeoff" (which, from his point of view, meant that the plane was lifting off). However, the controller misunderstood this seemingly unambiguous statement and inferred that the plane was actually on the runway, waiting to take off (i.e., it was stationary). Failing to understand that the "plane was accelerating towards take-off, a collision was all but inevitable" [62]. This fundamental issue (misunderstanding what another person says) must clearly be explained by any psychological theory of communication; yet, cognitivism found dealing with this sort of situation particularly difficult. The question of why this was the case brings us to the last of the four new psychologies: discursive psychology.

5.7 DISCURSIVE PSYCHOLOGY

Traditional cognitivist views of language tend to presuppose what George Lakoff and Mark Johnson termed the *conduit* (or pipeline view of language). In this view, items of information (conceptualized as discrete, quasi-physical entities) are retrieved from memory stores and put into language, which functions as a pipeline or conduit. The information units are then transmitted through language to the recipient's brain, where they are unpacked (presumably by a homunculus: certain, in a rule-bound fashion) and processed, after which the salient items are stored in memory stores, waiting to be transmitted to the next person [63].

It can be seen that the conduit view, is, as it were, the linguistic aspect of cognitivism. It presupposes memory stores and rule-bound cognition. Moreover, it views information as consisting of discrete units that map onto external reality. In other words, when I see a dog, then I map the word dog into it, and when I remember the dog, I put an internal representation of it into my memory store. This presupposes that there is a one-to-one relation between the internal representation of the dog and the real dog, and that this is an objective, context-free fact (anyone would see this relationship). Moreover, there is no ambiguity here. When I say "there is a dog," then this is transmitted directly to your head, and you know precisely what this means. But, recent experimental and empirical work has shown that this view of language is grossly misleading.

5.7.1 INDEXICALITY

Following from work in the hermeneutic tradition, the psychology of Vygotsky and (especially) the later work of Ludwig Wittgenstein, discursive psychology sees language not as a neutral pipeline that transmits thoughts to heads, but instead as a multifaceted thing that is used in many different ways and that needs to be interpreted actively to be understood.* But, what does it mean to understand language? The key point to understanding language is that it is always, intrinsically, indexical.

To state that a phrase or word is indexical means that its meaning changes according to context. For example, the word *bad* has a fairly set meaning. But, in some street slang, *bad* actually means good. Likewise, *yeah* usually means … well … yeah. However if said in a sarcastic tone of voice (or repeated) (dismissively) "yeah, yeah," it means "I don't agree." It is important to see that all language is like this. For example, in the example above, "We are now at takeoff" might mean "We have taken off" or "We are about to take off." In other words (to use jargon from hermeneutics), this phrase is polysemous: it can be interpreted in a number of ways according to context.

It is important to understand that (according to the hermeneutic tradition) all language is inherently polysemous, and it is impossible to say what a word or phrase means unless we know the context in which it is used. The context depends not only on the specific situation at the time (e.g., the tone of voice used), but also what Wittgenstein called the "forms of life," that is, the culture and environment in which we live. We need shared forms of life (e.g., the shared culture of a pilot) to understand language. For example, the language used by the pilots in the example is fairly opaque to outsiders, but it made sense to them.

Language is, therefore, situation specific (or contextual, if we remember the discussion of contextualism in Chapter 1). However, it is also something that is used actively rather than something that passively transmits information. People attempt to play whatever language game presents itself to them. Language use and discourse are best seen as motivated and (sometimes) an attempt to get things done rather than primarily veridical (i.e., an attempt to tell the unvarnished truth).

Not too much will be said about discursive psychology here, but it should be noted that it underlies at least part of taxonomy theory, the aspect dealing with language. Instead of viewing language as a pipeline (or window) through which meanings are passed, we can look at language as itself, as something that must be interpreted via the socially learned categories or taxonomic classifications we have learned. Moreover, it must always be remembered that describing something as human error is itself a kind of functional discourse. For example, no accident investigation will ever conclude that "we were unable to discover the cause of the accident due to our own incompetence. We are now going to use the phrase 'human error' to cover this up and blame the pilot or driver, who is conveniently dead and will be unable to answer our accusations." But, in reality it is significant that the phrase *human error* is usually used when all other possibilities are exhausted; in other words, it is a judgment by default. We could not find a mechanical cause; therefore, it "must have been" human error.

---

* Discursive psychology has been mainly developed recently by Jonathan Potter and others at the University of Loughborough.

5.7.2 FUNCTIONAL DISCOURSE

To pursue this point slightly further, when one gathers information from either a questionnaire or an interview, the temptation is always to imagine that one is getting at the cognitive facts that are passed through language from the heads of the people interviewed. But, we are better off seeing this as a systems relationship (for more details of systems, see the next chapter). In other words, any discourse produced is an emergent property of the interviewer/interviewee system.

There are numerous experimental studies that demonstrated that this is the case. For example, Davies and Baker [64] showed that, when interviewed by a fellow drug user, heroin users were more likely to stress the hedonic, pleasurable aspects of their experience, as opposed to when they were interviewed by a professional, for whom they were more inclined to state that they were addicted to the drug and that this was nonpleasurable and not under their control. There are other biases as well. For example, black children and white children give different results on IQ tests depending on whether they are given the tests by black or white testers [65]. These effects are not necessarily fatal to any question-naire/interview study, but they must be taken into account by triangulation (i.e., the same subject is interviewed/ given the questionnaire in two or more different settings and an average is taken of the results) or else controlled for (i.e., the same interviewer carries out all the interviews in as similar a setting as possible).* But, the effect cannot and should not be ignored. Again, discourse and results of tests are situated. What we get from the results of one test is precisely that: the results of one interview/test at a specific place with a specific person at a specific time. We extrapolate from this at our peril.

5.8 CONCLUSION

It should now be obvious why we believe that old-fashioned cognitivist psychology has not provided the insights it promised. It has pointed attention in the wrong direction: toward the alleged cognitive mechanisms, which generate error, instead of the social world in which we all actually live and work. According to Dekker and Hollnagel:

> The common way of thinking (i.e., in safety science) about performance implies that there is a set of internal or mental process that determines what people see and do ... (but) ... even if the information processing model of human behaviour was correct — which, by the way, it is not — this approach means that one is looking for a hypothetical intervening variable, rather than for the more manifest aspects of behav-iour. ... There is, however, an alternative, namely to focus on the characteristics of performance ... one point of favour of that is that we actually do have records of performance (but) ... we should not be overly concerned with the performance of (for example) the pilot per se, but rather of the pilot + aircraft, in other words, the joint pilot-aircraft system. [66]

---

* With nonconfidential reporting schemes, this is less of a problem. In CIRAS, we attempted to control for this as much as possible by having the same interview setting and questions.

Dekker and Hollnagel mention that the best way to do this is to talk about control, a subject we discuss in the next chapter and in Chapter 8 [66].

The new psychologies discussed here not only point attention away from the inner and toward the outer, but also point our direction, as Dekker and Hollnagel indicated, toward a look at the various systems in which we all live and work. For example, situated cognition sees a pilot as in the joint pilot-aircraft system, and distributed cognition and discursive psychology see us always embedded in social systems. But, what do we actually mean by the word system? It is this question that the next chapter discusses.

REFERENCES

1. Reason, J., *Human Error,* Cambridge University Press, Cambridge, U.K., 1990.
2. Zhang, J., Patel, V.L., Johnson, T.R., and Shortliffe, E.H., A cognitive taxonomy of medical errors, *Journal of Biomedical Informatics,* 37(3), 193, 2004.
3. Rasmussen, J., Pejtersen, A.M., and Schmidt, K. *Taxonomy for Cognitive Work Analysis,* Riso National Laboratory, Roskilde, Denmark, 1990.
4. Grant, S., Cognitive architecture for modelling human error in complex dynamic tasks, *Le Travail humain,* 60(4), 363, 1997.
5. Wagenaar, W.A. and van der Schrier, J., Accident analysis: The goal and how to get there, *Safety Science,* 26(1), 26, 1997.
6. Searle, J., Is the brain a digital computer? *Proceedings and Addresses of the American Philosophical Association,* 64(3), 21, 1990.
7. Dennett, D., *Consciousness Explained,* Penguin, London, 2004.
8. Gentner, D. and Stevens, A., *Mental Models,* Erlbaum, Hillsdale, NJ, 1983.
9. Johnson-Laird, P., *Mental Models,* Harvard University Press, Cambridge, MA, 1983.
10. Wetherick, N.E., More models just means more difficulty, *Behavioural and Brain Sciences,* 16(2), 367, 1993.
11. Simons, D.J. and Chabris, C.F., Gorillas in our midst: Sustained inattentional blindness for dynamic events, *Perception,* 28(9), 1059, 1999.
12. Fischer, E., Haines, R.F., and Price, T.A., *Cognitive Issues in Head-Up Displays* (NASA Technical Paper 1711), NASA Ames Research Center, Moffett Field, CA, 1980.
13. O'Regan, J.K. and Noë, A., A sensorimotor account of vision and visual consciousness, *Behavioral and Brain Sciences,* 24(5), 939, 2001.
14. Kotchoubey, B., About hens and eggs. Perception and action, psychology and neuroscience: A reply to Michaels (2000), *Ecological Psychology,* 13(22), 123, 2001. See also Kotchoubey, B., Event-related potentials, cognition, and behavior: A biological approach, *Brain and Behavioral Sciences,* under review (available online at http://www.uni-tuebingen.de/uni/tci/links/erp.pdf).
15. Gibson, J.J., *The Ecological Approach to Visual Perception,* Houghton Mifflin, Boston, 1979.
16. Glenberg, A.M., Symbol grounding and meaning: A comparison of high-dimensional and embodied theories of meaning, *Journal of Memory and Language,* 43(3), 379, 2000.
17. Pfeifer, R. and Verschure, P., Complete autonomous systems: A research strategy for cognitive science, in Dorffner, G., (Ed.), *Neural Networks and a New Artificial Intelligence,* International Thomson Computer Press, London, 1994.
18. Putnam, H., *The Threefold Cord: Mind, Body and World,* Columbia University Press, New York, 1999.
19. Atkinson, R.C. and Shiffrin, R.M., The control of short term memory, *Scientific American,* 225(2), 82, 1971.

20. Miller, G., The magical number seven, plus or minus two: Some limits on our capacity for processing information, *Psychological Review,* 63(2), 81, 1956.

21. Shallice, T. and Warrington, E. K., Independent functioning of verbal memory stores: A neurophysiological study, *Quarterly Journal of Experimental Psychology,* 22, 261, 1970.

22. Crowder, R., The demise of short-term memory, *Acta Psychologica,* 50(3), 291, 1982.

23. Reisberg, D., Rappaport, I., and O'Shaugnessy, M., Limits of working memory: The digit digit-span, *Journal of Experimental Psychology: Learning, Memory and Cognition,* 10, 203, 1984.

24. Glenberg, A.M., What memory is for, *Behavioral and Brain Sciences,* 20 (1), 1, 1997.

25. Craik, F.I.M. and Lockhart, R.S., Levels of processing: a framework for memory research, *Journal of Verbal Learning and Verbal Behavior,* 11, 671, 1972.

26. Wilcox, S. and Katz, S., A direct realist alternative to the traditional conception of memory, *Behaviorism,* 9, 227, 1981.

27. Bekerian, D.A. and Baddeley, A.D., Saturation advertising and the repetition effect, *Journal of Verbal Learning and Verbal Behaviour,* 19, 17, 1980.

28. Giuliana, A., Mazzoni, L., Loftus, E., and Kirsch., I. Changing beliefs about implausible autobiographical events, *Journal of Experimental Psychology: Applied,* 7(1), 51, 2001.

29. Mazzoni, G.A.L., Loftus, E.F., Seitz, A., and Lynn, S.J., Changing beliefs and memories through dream interpretation, *Applied Cognitive Psychology,* 13(2), 125, 1999.

30. Kelley, C., Amodio, D., and Lindsay, D.S., The Effects of "Diagnosis" and Memory Work on Memories of Handedness Shaping, paper presented at the International Conference on Memory, Padua, Italy, July 1996.

31. Nader, K., Memory traces unbound, *Traces in Neuroscience,* 26(2), 65, 2003.

32. Kotchoubey, About hens and eggs. Perception and action, psychology and neuroscience: a reply to Michaels (2000), *Ecological Psychology,* 13(2), 123, 2001.

33. Fodor, J., *The Modularity of Mind*, MIT Press, Cambridge, MA, 1983. However, see also Fodor, J., *The Mind Doesn't Work That Way,* MIT Press, London, 2000.

34. Spelke, E.S., Hirst, W., and Neisser, U., Skills of divided attention, *Cognition,* 4(3), 215, 1976.

35. Spelke, E.S., Hirst, W., and Neisser, U. Skills of divided attention, *Cognition,* 4(3), 229, 1976.

36. Hirst, W., Spelke, E., Reaves, C., Caharack, G., and Neisser, U., Dividing attention without alternation or automaticity, *Journal of Experimental Psychology: General,* 109(1), 98, 1980.

37. Gopher, D., The skill of attention control: acquisition and execution of attention strategies, in Meyer, D.E. and Kornbloom, S. (Eds.), *Attention and Performance XIV: Synergies in Experimental Psychology, Artificial Intelligence, and Cognitive Neuroscience,* MIT Press, Cambridge, MA, 1993, p. 299.

38. Ericsson, K.A., Krampe, R.T., and Tesch-Roemer, C., The role of deliberate practice in the acquisition of expert performance, *Psychological Review,* 100(3), 363, 1993.

39. Davies, J., Ross, A., Wallace, B., and Wright L., *Safety Management: A Qualitative Systems Approach,* Taylor & Francis, London, 2003.

40. Heidegger, M., *Being and Time*, trans. Macquarrie, J. and Robinson, E., Blackwell, Oxford, U.K., 1978.

41. Davies, J., Ross, A., Wallace, B., and Wright, L., *Safety Management: A Qualitative Systems Approach,* Taylor & Francis, London, 2003.

42. Darley, J. and Batson, D., From Jerusalem to Jericho: A study of situational and dispositional variables in helping behaviour, *Journal of Personality and Social Psychology,* 27(1), 100, 1973.

43. Haney, C., Banks, W.C., and Zimbardo, P.G., Interpersonal dynamics in a simulated prison, *International Journal of Criminology and Penology,* 1(1), 69, 1973.

44. Hartshorne, H. and May, M.A., *Studies in Deceit,* Macmillan, New York, 1928.

45. Adelstein, A.M., Accident proneness: A criticism of the concept based upon an analysis of hunter's accidents, *Journal of the Royal Statistical Society,* 115(3), 354, 1952.

46. Shaw, L. and Sichel, H., *Accident Proneness,* Pergamon Press, Oxford, U.K., 1971.

47. Wittgenstein, L., *Philosophical Investigations,* trans. Anscombe, G.E.M., Blackwell, Oxford, U.K., 1999.

48. Suchman, L.A., *Plans and Situated Actions,* Cambridge University Press, Cambridge, U.K., 1994, pp. 122–154.

49. Rauterberg, M., About faults, errors, and other dangerous things, in Ntuen, C. and Park, E. (Eds.), *Human Interaction with Complex Systems: Conceptual Principles and Design Practice,* Kluwer, The Hague, 1996, p. 291.

50. Grandjean, E., *Fitting the Task to the Man,* Taylor & Francis, London, 1971.

51. Brown, I., Highway hypnosis: Implications for road traffic researchers and practitioners, in Gale, A.G. (Ed.), *Vision in Vehicles III,* Elsevier, North Holland, 1991.

52. Callaway, E. and Dembo, D., Narrowed attention: A psychological phenomenon that accompanies a certain physiological change, *AMA Archives of Neurology and Psychiatry,* 79(1), 74, 1958.

53. Baddeley, A.D., Selective attention and performance in dangerous environments, *Journal of Human Performance in Extreme Environments,* 5, 86, 2000.

54. Rauterberg, M., About faults, errors, and other dangerous things, in Ntuen, C. and Park, E. (Eds), *Human Interaction with Complex Systems: Conceptual Principles and Design Practice,* Kluwer, Norwell, 1995, pp. 291–305.

55. Frese, M., Brodbeck, F.C., Heinbokel, T., Mooser, C., Schleiffenbaum, E., and Thiemann, P., Errors in training computer skills: On the positive function of errors, *Human-Computer Interaction,* 6(1), 77, 1991.

56. Weinberg, G. and Weinberg, D., *On the Design of Stable Systems,* Wiley, New York, 1979, p. 251.

57. Anderson, M., Embodied cognition: A field guide, *Artificial Intelligence,* 149(1), 91, 2003.

58. Brown, J.S., Collins, A., and Duguid, S., Situated cognition and the culture of learning. *Educational Researcher,* 18(1), 34, 1989.

59. Brown, J.S., Collins, A., and Duguid, S., Situated cognition and the culture of learning. *Educational Researcher,* 18(1), 35, 1989.

60. Woolfson, C., Foster, J., and Beck, M., *Paying for the Piper,* Thomson, London, 1997.

61. Hutchins, E. and Klausen, T., Distributed cognition in an airline cockpit, in Engestrom, Y. and Middleton, D. (Eds.), *Cognition and Communication at Work,* Cambridge University Press, Cambridge, 1996, pp.15–34.

62. Cushing, S., *Fatal Words,* University of Chicago Press, Chicago, 1997.

63. Lakoff, G. and Johnson, M., *Metaphors We Live By,* University of Chicago Press, Chicago, 2003.

64. Davies, J.B. and Baker, R., The impact of self-presentation and interviewer bias on self-reported heroin use, *British Journal of Addiction,* 82, 907, 1987.

65. Canaday, H.G., The intelligence of negro college students and parental occupation, *American Journal of Society,* 42, 388, 1936.

66. Dekker, S. and Hollnagel, E., Human factors and folk models, *Cognition Technology and Work,* 6(2), 79, 2004.

6 Cybernetics and Systems Theory

As we saw in Chapter 1, one of the key aspects of technical rationality (TR) is that any object or system can be broken down into small component parts, which can then be mathematically modeled in terms of their causal relationships to one another. For a small class of events, TR works fine. The problem is that for extremely complex systems (and in safety, we are almost always faced with such systems) we may get a false simplicity; in other words, our little models with the little boxes and the little arrows may be so straightforward as to be somewhat crude and therefore misleading. To say otherwise is to "mistake the map for the territory." Luckily, there is another approach that looks at systems as a whole and studies the component parts of systems insofar as they contribute to the whole: systems theory.

So, what is systems thinking or systems theory, and what is its relevance in the context of this book? The answers to those questions depend on what we mean by a system. From the point of view of systems theory, the answer, luckily, is fairly simple. As the systems theorist Peter Checkland put it: "The central concept 'system' embodies the idea of a set of elements connected together which form a whole, this showing properties which are properties of the whole, rather than properties of its component parts" [1]. In other words, systems theory analyzes complex phenomena that exhibit the property of emergence, which we discussed briefly in Chapter 1.

The next question is, what is emergence? Checkland used the example of water: water's wetness is a feature of the two molecules of hydrogen and one of oxygen, but only when they are brought together. In other words, water's wetness is a macro feature, distinct from the microstructure of the individual molecules. Another example might be the color green. If one mixes together blue paint and yellow paint, we get green paint, but there are no green elements in the yellow or blue. The green is an emergent property of the yellow and blue.

In a sense, these are bad examples because they discuss only two elements (hydrogen and oxygen, blue and yellow). In real life, emergent properties are frequently the product of dozens or even hundreds or thousands of elements, which, together with the unpredicted emergent properties, make the resultant creation function as a system [2].

Systems theory was developed in the 1920s and 1930s but only really came to the fore after World War II, with the writings of researchers such as Ludwig von Bertalanffy, whose work *General Systems Theory* might well be considered the Bible of the movement [3]. Von Bertalanffy was a biologist, but it was in electronics and mechanical engineering that systems theory really caught on in the 1950s and 1960s, and the reason was that World War II had made everyone aware of the power of complex systems. After all, what was the war itself but an incredibly complex system involving millions of intermeshed people, machinery, and electronics?

It was also after World War II that the new science of cybernetics (from the Greek *Kybernetes* for "steersman") was created: the science of communication and control. Its origins lay in the steam governor developed by James Watt in the 18th century. Watt's governor was important because it used feedback to self-regulate (concepts described in Section 6.2). However, it was the American scientist and engineer Norbert Wiener who developed the modern form of the science based on his work with machine gunners and their attempts to shoot down enemy planes [4].

So, what is cybernetics? Normally, a description is given in terms of the cruise control mode in a car or of a thermostat, but this unfortunately gives the impression that cybernetics has something *necessarily* to do with electronics or computing (although in the real world it normally does). So, let us choose a more homely example: the modern flush toilet. According to the American psychologist Gary Cziko:

> The modern flush toilet must have a certain amount of water on hand for each flush to be effective. For this purpose, most residential toilets make use of a holding tank into which water accumulates between flushes. Since too little water in the tank does not allow adequate flushing and too much is wasteful (it will simply flow out through an overflow drain), a mechanism is used to maintain the water at the desired level. This mechanism consists of a float resting on the surface of the water that is connected to a valve. When the water level falls after a flush, the float falls with it and in so doing opens a valve, admitting water into the tank. But as the tank fills and the water level rises, so does the float, eventually closing the valve so that the tank does not overfill.

> For the reader who has not already peered inside a flush toilet tank, it is well worth lifting the lid and taking a look. With the tank lid off and the flush lever activated, one can observe in live action the events described: the tank empties, the float falls, the valve turns on, the tank refills, and the valve shuts off. It is also informative to push lightly on the flush lever for a few seconds so that just a portion of the water in the tank escapes into the bowl. This will show that the tank need not be emptied completely before the float-valve mechanism acts to refill the tank. If all is operating properly, the float-valve mechanism will not let the water remain very much below the desired level. What is this desired level? Inside most tanks a line indicates the optimal amount of water for flushing the toilet. If the water level in your tank is above or below this line, it can be changed by adjusting the float's position on the link that connects it to the valve. By changing the distance between the float and the valve, you can control the water level that will be reached before the valve turns itself off. [5]

Now, what is interesting about a humble flush toilet? First, if one thinks in terms of information transfer, what is happening here is that information is transferred from the water level to the valve, such that the valve will act. However, because of the float, information is also transferred *back* to the water level to ensure that the tank does not overflow. This is the phenomenon of feedback familiar to most people from the experience of holding a phone or microphone too close to a compact disk player or an electric guitar too close to an amplifier. When this happens, the instrument feeds back, which produces the high-pitched squealing noise that makes most of us separate the two things as quickly as possible.

That is an example of undesirable feedback, which we generally want to avoid. But, note in the toilet example above that the information flow (and water flow) is a *good* thing: the feedback of information helps the system (i.e., the toilet) to self-regulate and therefore not flood the bathroom floor.

We can go further than this, however, and imagine positive and negative feedback. A classic example of positive feedback is an arms race (normally found in the natural world* but still, alas, also found in the human world). For example, before World War I the German and English fleets ran a race to see who could create the world's most powerful naval destroyer. Whenever information (i.e., feedback) got back to the Germans that the English had just built a bigger ship, they felt duty bound to produce one that was even bigger; so, with this positive feedback, ships simply got bigger and bigger and bigger. The example of the relative nuclear arsenals of the West and the East after World War II is another classic example of an arms race employing positive feedback.

But, ipso facto, there must also be such as thing as *negative* feedback, by which a negative response would lead to a smaller action, which would lead to an even more negative response, and so on. Modern amplifiers use this technique to reduce feedback (preferably to nothing).

What is the aim of most cybernetic systems? In most cases, it is not to increase their power infinitely (i.e., using only positive feedback) or to reduce it to nothing (i.e., using only negative feedback) but instead to achieve what physicists term (in another context) the Goldilocks point, that is, not too cold, not too hot. In the toilet example, this is when the water level in the tank is at the midpoint, with just enough water to flush but not so much that it overflows. In cybernetics, this is termed the point of *homeostasis*.

So, cybernetic systems use feedback, both positive and negative, to attain homeostasis, and their ability to do this depends on the amount of control they have, a word that in cybernetics has a very specific sense: it refers to the extent to which the system can *self-regulate*.

There is one more feature of cybernetic systems that is important; this time, unfortunately, we *have* to use the example of a thermostat. A thermostat attempts to control the temperature of the room, again to achieve homeostasis (not too hot, not too cold) against buffering from outside the system. But, imagine that a smart engineer decided to help the thermostat by installing a sensor outside the house, such that when cold weather (or unusually hot weather) was on the way the thermostat could plan ahead and put the boiler on to produce greater reserves of heat. This is feedforward and enables the system to be proactive in terms of threats to homeostasis.

To recap, in a cybernetic system there are constant feedback and feedforward loops of information, with the controller attempting to self-regulate the system. So, a thermostat uses feedback (from temperature sensors) to regulate the temperature of the room; cruise control uses feedback to regulate the speed of a car; and so on. The key point is control (cybernetics is sometimes termed the science of control): how much control the controller has over the system. The more control it has, the more efficiently it will be able to accomplish its purpose.

---

* For example, this is found in the relationship between a predator and its prey.

Second, now let us be reminded briefly of what Cziko said in the section concerning the toilet, especially the last sentence. Later, Cziko elaborated on this statement:

> Notice the phrase I used in the preceding sentence — "the valve turns itself off." Is this actually the case? Isn't it rather that the rising float causes the valve to close? Yes, of course. But what is it that causes the float to rise? Obviously, the water that is filling the tank. And why is the water entering the tank? Because the valve is open. And what will cause the valve to close? The rising water level. So the valve, through a series of events, does in a sense close itself, since the valve's opening eventually causes it to close again. If it seems that we are going around in a circle here, it is because we are. [6]

This is what in cybernetics is termed *circular causality* in which things *cause themselves* to happen. As the example of the toilet shows, there is absolutely nothing mysterious or mystical about it; nonetheless, it shows two things: the one-way model (or metaphor [7]) of causality is absurdly misleading when applied to complex systems (instead, it is better to talk of systems going in and out of control) and that even inert, nonelectronic, nonbiological systems can exhibit teleology or, in layperson's terms, purpose. The valve causes itself to do certain things, and therefore its (self-derived, in this particular context) purpose is to do these things. Certainly, in this case it is what philosophers term *derived* teleology (the valve did not build itself, someone in a factory did), but it is teleology nonetheless. The implications of this are discussed below.

Third, what is the link between cybernetics and systems theory? The link is that it is impossible to talk about the purpose of the valve without discussing the rest of the toilet mechanism; it functions, therefore, as a complete system. To emphasize a distinction made earlier, this is not an ontological point; you could certainly break down or reduce this system to its component parts if you wanted. But, you would be none the wiser about how it worked: about its teleology or purpose, about how it functions dynamically. To do this, you have to look at the whole system because the system as a whole displays emergent properties, properties related to its seeking after homeostasis and its manifestation of control.

6.1 SECOND-ORDER CYBERNETICS

During the 1950s and the 1960s, cybernetics and systems theory concentrated mainly on technical systems. However, by the 1970s scientists were starting to wonder if biological systems might also be regarded as, in some senses, control mechanisms. To see why this is the case, we have to look at two other features of systems.

We have said that the toilet system attempted to achieve homeostasis. But, it does not have to. In the context of a thermostat, we can imagine an electronic mechanism that continually varied the desired temperature such that the goal of the system was perpetually changing. This is termed *rheostasis*, a homeostatic point that varies over time. However, other points that were related to the change in systems theory and cybernetics were more radical.

In what was termed *second-order cybernetics*, which began to be developed in the 1970s by scientists such as Heinz von Foerster, it was argued that living beings were indeed cybernetic systems, endlessly changing rheostasis in a dynamic and changing world. However, if that was the case, then human beings were also cybernetic systems; therefore, this was a case of cybernetic systems looking at other cybernetic systems (hence the phrase cybernetics of cybernetics, often used to discuss second-order cybernetics). Therefore, our looking at these systems was not an unmotivated act but was instead part of our own feedback and feedforward mechanisms [8].

This fitted in with the philosophical view termed *perspectivism* (i.e., the view that when we see something we always see it from a specific perspective), which of course is extremely similar to the contextual views we have looked at elsewhere (we discussed both these views briefly in Chapter 1). In the language of systems theory, we can explain it this way: when we look at something (even more when we interact with it), we create a new system with new emergent properties. To ask what the system was like before we started to do this is meaningless [9].

This ends the discussion of epistemology. But, as systems theorists began to look at biological and then sociological systems (this new approach was termed *soft systems analysis*), new concepts had to be developed to analyze the systems property of living beings. These new concepts are morphogenesis and autopoiesis.

Morphogenesis refers to biological systems (e.g., a fertilized human egg) that tend to increase the complexity of their internal structure over time. *Autopoiesis*, again, refers to a system that reproduces itself (e.g., a virus or DNA). These systems evolved via the brute force of natural selection, but because of these properties, they manifest teleology: purpose [10].

6.2 CYBERNETICS, SYSTEMS THEORY, AND HUMAN BEHAVIOR

So far so good, but we are left with the following problem: what relevance does all this have to safety? We deal with the specifics of that issue later, but the key point to understand here is that these biological (and social) systems manifest *self-organization*. What is it from outside the fetus (or child) that enables it to grow? Certainly, it needs energy and stimulus from outside, but it sucks these out of the surrounding environment rather than passively awaiting for them, as traditional stimulus–response or stimulus–cognition–response theories would have us believe. This is the great move forward in understanding via complexity theory, the science of self-organizing systems (which is itself a subbranch of systems theory) [11].

Cybernetics and systems theory, despite their different emphases and origins, are normally considered part of the same science because they are philosophically opposed to reductionism, the idea that systems can always be understood by breaking them down into their component parts and looking at simple causal relations between them.

Now, so far we have simply looked at cybernetic/systems models as abstract mathematical models of the way the world can be viewed. But, second-order cybernetics goes further and states that human beings *do actually behave* like cybernetic control systems. Is this in fact the case?

6.2.1 P3 Waves and the Sensorimotor View of Perception and Action

To see whether this is the case, we have to go back to the findings of neuropsychology. Again, we have to remember that, although cybernetics was discovered after World War II, it had precursors, for example, the work of American psychologist and philosopher John Dewey. In his classic essay, "The Reflex Arc Concept in Psychology" (1896), Dewey argued against what was then becoming the orthodoxy (stimulus–response theory) in favor of something he called the reflex arc, which now looks very similar to the classic cybernetic feedback loop [12]. The basic difference, of course, is that stimulus–response and stimulus–cognition–response both view the organism as basically passive. The reflex arc and cybernetic views both see the organism as fundamentally active and teleological, that is, having goals.

Boris Kotchoubey (whose work we discussed in the last chapter) has carried out experiments to test the theory of the reflex arc [13]. Modern monitoring equipment can detect the P3 brain wave, which is associated with higher-level cognitive thought. If the cognitivist/behaviorist view is correct, then when we are given a stimulus there should be a very brief pause, then a cognition (i.e., the P3 wave), then the action. In other words, in this view the organism will be passive (standing around on a street corner or whatever) until someone comes along and does something, say, like punching the person in the face (stimulus). There will then be a pause, then a cognition (while the organism processes this information) that will be associated with the P3 wave, then a response (running away, yelling, punching the opponent back, or whatever). So, does this actually happen?

The answer is simple: no. Kotchoubey found numerous situations when the P3 waves (i.e., the cognition) came *before* the stimulus. Needless to say, this has been noted in previous experiments and is usually explained as a preattentive cognition. But, Kotchoubey argued strongly that this view makes no sense: imagine there *was* no stimulus. Then, what precisely is pre about this attentiveness? Kotchoubey argued that this is simply playing with words to rescue the stimulus–cognition–response orthodoxy. The fact is that the brain here is simply active and does not need an external stimulus (when you open your eyes in the morning and get up to make a cup of coffee, to what stimulus are you reacting?). The language of cybernetics (where toilets, thermostats, and people cause themselves to do things) seems far more appropriate to describe this state of affairs [13].

But, if this view of circular causality is accepted, then we are led to accept the other corollaries of cybernetics, what philosophers term the breakdown of the subject-object distinction. Remember the behavior of a thermostat (subject) cannot be understood except with reference to the temperature of a room (object). In other words, the thermostat and the room constitute a thermostat-room *system* (to use the language of fuzzy logic, the distinction between subject and object is an analog one, not a digital one; but, for fundamentally analog phenomena, we often make taxonomic distinctions, in this case between subject and object).

In the same way, if we examine a human (i.e., a control mechanism, subject) driving a car (object), then it is clear that this is a human-car system. We can now link to situated cognition or what Heidegger attempted to explain as being-in-the-world. It does not really matter whether we are using the language of Heidegger's phenomenological hermeneutics or that of cybernetics or systems theory, the key point made is the same: there is no such thing as an abstract human floating in metaphysical space. Human beings are always *doing* things (active, not passive), and they are always in *a context* (whether a car, a plane, a supermarket, a desert, or whatever).

This aspect of cybernetics and systems theory and its implications for safety have been grasped most strongly by Erik Hollnagel, who has described the driver in the car thus: "The model [i.e., the cybernetic model of driver behavior] describes driving in terms of multiple, simultaneous *control loops* with the Joint Driver-Vehicle System (JVDS) as a unit" [14; italics added]. In other words, as we saw in the previous chapter, when we see a driver in a car, we have to think not of two things but one. By the same process of logic:

> The pilot-plane ensemble (also) clearly constitutes a JCS (Joint Cognitive System), which exists within the larger air traffic environment. It makes sense to consider the pilot-plane ensemble as a JCS and to analyze, e.g., how well it is able to attain its goals, such as getting to the destination at the scheduled time. Alternatively, the pilots in the cockpit can also be considered as a JCS that is studied separately, such as in Cockpit or Crew Resource Management. The environment of the pilot-plane or crew-plane ensemble can be described in terms of the air traffic management, the system of flight routes and traffic sectors, and the weather. The "natural" boundary depends on the purpose of the description, whether it is crew resource management or free routes. It follows that the position of the boundary cannot be established by formal means alone, but that it must be done *pragmatically*. [15; italics added]

From this point of view, we can see the pilot-cockpit system as a unified control system, but we could also broaden our view of the situation to include the air traffic control system, the external environment, or anything else, and the criteria for what system we should look at depends on what we want to find out. This is why cybernetics and systems theory are sometimes viewed as the same theory: in both cases, the focus of analysis moves from the individual to the system. It should be clear that this is a radically different approach from the cognitivist approach, which, on the contrary, moves deeper and deeper into the individual in an attempt to find internal cognitive modules that result in human error.

6.3 CYBERNETICS: CONCLUSION

To repeat, at this point we have looked at cybernetics (especially second-order cybernetics, which deals specifically with biological systems) and systems theory. There is a considerable amount of evidence that humans are in fact cybernetic control mechanisms. Therefore, when we see a driver, or when we see a pilot in a cockpit, we can see them as driver-car systems (joint cognitive systems, as Hollnagel called them) or a pilot-cockpit system. The pilot/driver is attempting to self-regulate the system and prevent it from going out of control. When a driver drives, he or she is using feedforward loops (which might come from the radio, the map, or previous knowledge) and feedback loops from vision, hearing, and touch to self-regulate the car-driver system and achieve the ever-shifting rheostatic point, where the car is in the right place. When something goes wrong (e.g., the car hits an oil slick), the feedback becomes misleading or inadequate, and the system, literally, goes out of control, and an accident occurs.

But, as the presence of the oil slick shows and as Hollnagel's example demonstrates, the size of the system we are looking at is primarily a pragmatic decision. In the case of a plane, we might decide that the pilot-plane system was the most relevant one or that the pilot-copilot-plane system or the pilot-copilot-plane-air traffic controller system was the most useful unit of analysis.

6.3.1 OPEN AND CLOSED SYSTEMS

This distinction of Hollnagel's is similar to the standard taxonomic classification of systems as open or closed. Flach emphasized the key point of second-order cybernetics (and soft systems methodology):

> A system is not a physical entity that exists independently of an observer (e.g., a researcher or analyst). ... A system is an understanding about how physical components are connected ... keeping this in mind, the distinction between open and closed systems will be defined in terms of what the analyst knows. ... A closed system is defined as a system where all the components and all the rules for interaction are understood by the analyst. ... Examples of closed systems are games such as tic-tac-toe, checkers or chess. For each of these systems it is possible, in principle, for the analyst to enumerate all possible inputs, states and outputs. An open system is defined as a system that exceeds the analyst's understanding. ... All human socio-technical systems are open systems. [16]

So far, we have looked at systems that, although technically open, are small scale enough to be considered closed for some purposes. However, we should now look at much larger systems that are open both in theory and in practice; that is, they inherently contain interactions and processes that are unknown and unsuspected to the analyst — and the user.

6.4 NORMAL ACCIDENTS

To be clear, anything can be viewed as a system: an organization, a company, anything. Cybernetics seems be backed up by experimental data and seems to provide a useful way of looking at real-world safety situations. What about systems theory? Does it actually provide worthwhile and meaningful safety interventions in the real world? To answer this question, we pursue a slightly unusual course. We look at Charles Perrow's normal accidents theory (NAT). Because that is a systems theory approach, if we can demonstrate that NAT makes meaningful predictions and leads to effective interventions, then we will assume that systems theory has some validity (in other words, as NAT is a subcategory of the supercategory systems theory, if NAT works, then systems theory must by definition also work to some extent).

6.4.1 Systems Accidents and Normal Accidents

NAT is a theory that was first formulated in the book *Normal Accidents* by the sociologist Charles Perrow [17]. Perrow's basic argument is that certain kinds of accidents (systems accidents) are *literally inevitable* in certain systems; hence, the phrase *normal accidents*. It is *irrelevant* how many safety nets and safeguards are set up; certain kinds of accidents will always happen. In fact, some of the safeguards put in place may actually make these systems accidents more likely. Therefore, it is only by looking at the whole system (not the individual) that we can make meaningful predictions regarding where accidents will happen (although *when* they will happen and *how* they will happen will forever be unpredictable).

Some people have inferred that Perrow stated that *all* systems will create normal accidents, but this is not the case. Instead, Perrow argued that only systems that exhibit two specific features will create normal accidents. The first of these features is tight coupling.

6.4.2 Tight Coupling

According to Perrow: "Tight coupling is a technical term meaning that there is no slack or buffer or give between two items. What happens in one directly affects what happens in another" [18]. In other words, tight coupling describes the way in which two phenomena interact. We deal with tightly coupled systems all the time. When we switch the light switch on, the light goes on automatically (and, seemingly, instantaneously). If we set the burglar alarm and then walk into the beam, the alarm goes off automatically. It does, technically, take some time between the time when I turn on the light switch and when the light actually goes on, but this time is far too short for me to act. Perrow had a list of features that characterize tight coupling, such as the fact that delays in processing are not possible (there is no temporal gap, or so it seems, between flicking the light switch and the light going on), there is only one method to achieve the goal in principle (the only way to get the light bulb on is to switch the switch), and any buffers that are available were deliberately planned and put in (e.g., a fuse box in the light example) [19].

6.4.3 Nonlinear Systems

The second feature of the sort of system we are talking about here is that it be nonlinear or complexly interactive. This mysterious phrase is merely a way of talking about systems of the sort we discussed earlier in the chapter, for which the simple cause-and-effect metaphor breaks down. Nonlinearity has a technical definition in mathematics, but what is meant here is merely a system in which simple reductionism breaks down.* A system that exhibits *emergent properties* is nonlinear and therefore

---

* Perrow did not use precisely the mathematical definition of nonlinear and neither are we, but the two definitions are close enough to be used as synonyms here.

will be at the mercy of the feedback and feedforward loops. The effects of these properties are well known from the examples of chaos and complexity theory that have made modern systems theory famous. In systems like this, small phenomena might have large effects, or large phenomena might lead to little or nothing. Moreover, there might be inexplicable feedback effects that achieve the opposite of what we were trying to achieve in the first place (blowback). The key point is that these effects are unexpected and, in a fundamental way, unpredictable. The system has emergent properties that could not be inferred from an analysis of its component parts or predicted even by the engineers who built the system. Complex or nonlinear systems are characterized by interconnected subsystems, feedback (and feedforward) loops, multiple and interacting control systems, and an incomplete understanding by any one individual (even an expert) regarding how the system really works. The classic example of a system that manifests nonlinearity and tight coupling is a nuclear power plant, and Perrow therefore predicted that normal accidents are inevitable in nuclear power plants, no matter how many safeguards are put in place.*

6.4.4 SPACE FLIGHT AS AN OPEN SYSTEM

Perhaps an even better example of a complex and tightly coupled system is space flight, which we look at for two reasons: first, it is a good example of normal accidents in practice; second, it illustrates a way of getting out of these normal accidents. The classic example is the now-famous flight of *Apollo 13* (an example of a normal accident with a less-happy conclusion is the *Challenger* disaster, which we examine in the next chapter). This is a good example of Perrow's approach because the *Apollo* flights in general and *Apollo 13* in particular showed features of a highly nonlinear and tightly coupled astronaut-spacecraft system. Therefore, according to Perrow, accidents were inevitable. However, *Apollo 13* is a particularly good example because, even though the initial accident was a classic normal accident, disaster (i.e., loss of life) was eventually avoided by putting the operators back in the driving seat (i.e., giving them back their control) as well as moving back to a more linear and loosely coupled system. This enabled the astronauts to control and understand their system more clearly and therefore recover the mission (as much as was possible). The next section draws heavily on Perrow's discussion of *Apollo 13*, to which we would also refer readers (see Chapter 8 of [17]).

6.4.4.1 *Apollo 13*

First, some background regarding *Apollo 13* is necessary. In the 1950s and 1960s, the engineers at the National Aeronautics and Space Administration (NASA) did everything they could to create a spacecraft that would work without human

---

* Just to clarify, the nonlinearity and tightly coupled taxonomic distinctions have nothing to do with the taxonomic distinction between an open and a 'closed' system. If a system is tightly coupled, then it means that there will be an (extremely) short period of time between interactions in one part of the system and interactions in another. Nonlinearity, on the other hand, refers to systems that manifest emergence. To say that a system is open or closed, on the other hand, merely refers to the amount of knowledge we have of that system. Most systems of the sort discussed here (i.e., high-consequence, man-machine systems) should be considered open unless this can be proved otherwise.

operators (this is a logical deduction if you believe that there is something called human error: if you get rid of the human, then you get rid of the error).

The very term *astronaut*, although it sounds glamorous, was originally intended to make explicit that humans just hitched a ride on the early American space ships. The original term, of course, was *pilot*, but this was felt to have implications of control and agency that were out of place. Instead, astronaut (literally star voyager) was coined as a more *passive* term. At one point, it was seriously argued that all the astronauts should be given tranquilizers throughout the voyage in case they touched anything or in some other way committed a human error. Engineers (behind the astronauts' backs) described them as a redundant component and complained that astronauts were only allowed to be in the spacecraft for political reasons (which was undeniably true).

What caused the *Apollo 13* near-disaster? As always, it is difficult to say (see Chapter 2). However, a plausible sequence of events that was reconstructed afterward follows.

Two weeks before the lift off, during a routine prelaunch test, the oxygen tanks of the spacecraft were tested. These carried not only liquid oxygen (which was kept at close to absolute zero temperature), but also the elements for fuel cells that generated the electricity for the craft. At the end of the task, because they had difficulty getting the oxygen out of the tanks, the engineers turned on the tank heaters, but then seem to have forgotten about them and left them on. One of the main reasons they took little notice of the status of the heaters was that there was an automatic sensor that normally turned the heaters off when the safe temperature range was exceeded. But, this sensor was designed to work in outer space, that is, in situations of low power and low temperature. What no one had predicted was that, in situations of *normal* power and temperature, the sensor did not work properly. So, the temperature continued to climb until the heat burned out all the insulation for the electric wires as they entered the tank. Note that this happened *after* all the electrics had passed their final inspection, and there was no other automatic sensor to indicate there was a problem.

This brings us to our first and most important point: even though normal accidents are inevitable, they are also unpredictable. We can know that they will happen but *never* how or when. This is not only because of the arguments against determinism we saw in Chapter 1, but also because of the very nature of nonlinear systems: it had simply never occurred to anyone that the sensor might fail in this particular way. Open systems have internal relationships between components and processes that are not conceived of by the original designers. It is simply impossible to think of all possible interactions.

Therefore, the spacecraft went into space with the key insulation for important electric component wires burned away. The first trouble the astronauts noticed was when the now-uncovered wires short-circuited, causing an arc, which eventually heated the oxygen tank until the cap blew off. This resulted in one of the hatches of the craft blowing off in a small explosion. This explosion was so strong that it was felt aboard the craft. The arc then caused a fire.

Oxygen was now venting from the spacecraft, but the short circuit meant that there was also a loss of electrical power. The venting oxygen was also disturbing the attitude of the craft. Communication with ground control was also becoming intermittent (as wild variation in the attitude of the craft, caused by the venting oxygen, meant that the radio antenna found it difficult to lock onto the Earth transmission).

This situation, of course, led to various warning lights and alarms going off, one of which (the oxygen quantity gauge) indicated the nature of the problem. However, what was missing was a meta-alarm to point to the *significance* and *meaning* of the alarms. In retrospect, the oxygen alarm was the key one, but with many other alarms going off, no one knew this at the time; in fact, the number of alarms indicated more strongly that there was a problem with the alarm system.

When the astronauts managed to regain contact with mission control, they explained their position. However, here we run into yet another fundamental problem. Humans (even the humans who actually built the system) have little intuitive grasp of nonlinear systems. We are all happier in the world of simple cause and effect. Because the engineers had built the system and had (they imagined) thought of all eventualities, the possibility that there might be a serious problem was suppressed. Amazingly, whereas in earlier missions astronauts had been forced to go through seemingly implausible *systems* failures in the flight simulator, by the time of *Apollo 13* failure scenarios were restricted to simple component failures. The possibility of failures of emergence in which the failure is an emergent property of three or more failures (as was the case here) simply seemed to be too implausible and time-wasting.

Therefore, the damage continued to get worse as mission control continued to tell themselves that nothing could be seriously wrong. They were helped in this supposition by the fact that the alarms were giving misleading information and sometimes contradicted one another. After a few minutes, almost half the warning lights on the control monitor had lit up. Which ones indicated the real problem, and which were a distraction? Which did not really indicate a problem at all but had been triggered by other electrical problems? There was no way of knowing.

Eventually, the crew was reduced to reading aloud all the gauges in the command monitor that had anything to do with electrics. When they got to the oxygen pressure gauge, they saw that it was indicating that the tank was empty. Baffled by this, one of the astronauts happened to look out the window and actually saw the oxygen streaming out of the craft (of course, they had previously felt the bang of the initial oxygen burst but had not understood what this was). This gave them the crucial insight they needed to solve the problem.

This brings us to another key point that backs up the view of human psychology discussed in the last chapter: the difference between situational, embodied knowledge, and disembodied knowledge possessed by experts. Despite the fact that the ground controllers had vastly more computer power (and manpower) available and despite the fact that they had actually *built* the spaceship, at this point they still knew less about the situation than the astronauts. The reason why is simple: it is to do with the embodied nature of the astronauts' knowledge. The astronauts had *felt* the bang as the oxygen began to blow into space; the ground controllers had not. The astronauts *saw* the gas leaking into space; the ground controllers could not. The astronauts were situated within the spaceship; the ground controllers were far away from it, and their knowledge was disembodied, abstract. Moreover, they were also the victims of a double bind: *because* they were experts and *because* they had built the craft, they persuaded themselves that a complex systems failure of this sort simply could not be happening. So, valuable time was lost. At this point, the astronauts, not the controllers, were the experts, despite the controller's greater technical knowledge.

There is no need to go into great detail here about the now-famous recovery of the *Apollo 13* mission. It is sufficient to say the three astronauts eventually had to spend four days living in the lunar module (which had never been designed for such an eventuality) to preserve oxygen and power before moving back to the command module (the lunar module had no heat shield).

What is interesting here is *what had to be done to save the mission in safety science terms*. The spaceship-control system would have to go from a nonlinear system to a linear system (as much as possible), from a tightly coupled system to a loosely coupled one. All extraneous (and therefore nonlinear) features of the mission were discarded. Instead, there was now to be only one goal: getting the astronauts back alive. Reentry procedures were drastically simplified. Controllers had to pick options that decoupled systems such that they could not become out of control and trigger other systems without the controllers' (and astronauts') knowledge. Most important, the controllers had to reactivate the astronaut-craft control system, which had been deemphasized in favor of the controller-craft control system. The astronauts were no longer passive objects in the system; they were forced to become active subjects. They were forced to regain control.

6.4.5 CONTROVERSIES AND ARGUMENTS

Perrow's thesis was and is extremely controversial because it challenges many Western biases pertaining to causality and how people and machines interact. In the Western educational system, we are all taught not only facts and values, but also styles of explanation. We are taught that the best way to analyze a system is to break it into its component parts (reductionism) and to look at the simple causal connections between these parts. Moreover, as we saw in Chapter 1, we are encouraged to have an internal locus of attribution: to emphasize our own control (and knowledge) of events. "Knowledge is power" remarked Francis Bacon many centuries ago; we become uneasy when are told that, for open systems, it is *impossible* to have total knowledge of them, and that therefore our power over them will always be limited.

It would be superfluous here to quote from the many safety pamphlets, leaflets, and books that state that, in theory if not in practice, *all* accidents are predictable, and that therefore *all* accidents are preventable. Normal accidents theory states that both these claims are false. We can predict, to be sure, that some systems will inevitably lead to accidents, but, to repeat, we can never predict when or how these accidents will happen. Moreover, although we can alter the system to a certain extent, our power (and knowledge) here is also limited: it seems likely that some high technology industries simply *have* to be tightly coupled and nonlinear, at least to a certain extent (space travel and nuclear power plants are excellent examples).

Therefore, these industries will always have accidents regardless of how many safety nets we install.

6.4.6 Nuclear Defense as a Normal Accident-Producing System

If NAT holds, it has disturbing implications for safety practice. We can study its applicability further by looking at Scott Sagan's analysis of the nuclear weapons defense system in the United States, a work that examines this specific issue [20].

Sagan began by dealing with a rather obvious problem for NAT. If there is any system that meets the criteria of nonlinearity and tight coupling, it is the elaborate system of nuclear missiles that makes up the American nuclear defense shield. But, so far at least, there have been no major nuclear incidents that have happened by accident. Why not?

Sagan began by contrasting two alternative philosophies of system accidents: what he called high reliability theory and NAT. High reliability theory proposes that with adequate procedures, safeguards, and design safety nets, accidents can be, for all intents and purposes, avoided. This school argues that the empirical record of such systems as the Federal Aviation Administration air traffic control system, the Pacific Gas companies' electric power system, and some (not all) of the peacetime operations of two specific American aircraft carriers shows that accidents can be, for all intents and purposes, prevented forever.

High-reliability theorists (e.g., Aaron Wildavsky) argue that, to make a system safe, safety must be taken seriously throughout the whole system (which means that it must begin by being taken seriously by management). The reason why most industrial systems are not safe, according to this school, is mainly political: safety costs money, and in a world of limited resources, most firms and organizations are simply not prepared to spend the money on genuine safety improvements.

Taking safety seriously means that there must be clear and easily accessible procedures, and there must be a good safety culture in which such procedures are widely followed. There must also be clear communication from management to staff and from staff to management. Moreover, there must also be redundancy, which means there are numerous safety nets so that if one system fails, another will quickly catch the error before it can become more serious. So, in high-reliability systems there will frequently be duplication (two or more systems to perform the same operation) so that if one system fails, another will kick in automatically. This functions with personnel as well: there must be a culture of responsibility such that people take responsibility for their own actions and can improvise (safely) if procedures fail or they meet some unforeseen eventuality. This is emphasized by constant safety training. Finally, a high-reliability organization will be able to *learn from its mistakes*. If by chance something does go wrong, personnel will sit down and work out why it went wrong so that this eventuality can never happen again; the knowledge gained will then be reproduced via training and new procedures. Over time, the system will tend to become safer and safer until accidents more or less drop away to zero.

6.4.6.1   High Reliability and Normal Accidents
Theory Compared

The alternative to high reliability theory is NAT, but Sagan's own version of the latter broadens it and introduces some important new concepts. The most important of these is the garbage can model of organizations pioneered by Michael Cohen and Johan Olsen [21]. This (deliberately provocative) title emphasizes that organizations rarely, if ever, function as they "ought" to according to abstract laws of logic or the abstract laws of rationality as used in, for example, economics. Instead, organizations can have unclear and inconsistent objectives, can be un-understandable (even by people within the organization), and can manifest fluid participation in their information transmission processes (in other words, some people go to meetings; others do not. Some people pay attention; some do not. Some people know what they are talking about; others do not. We saw this effect, on a microlevel, in the discussion of distributed cognition in the Chapter 5). Therefore, there are fundamental political and sociological complexities that mean that organizations will *never* function as they claim in abstract managerial mathematical models.

This version of NAT obviously makes different predictions from high reliability theory. To begin, NAT stresses that an organization cannot possibly have only one goal. Even if one puts safety as the highest goal, there will always be other aims and objectives (production is the most obvious one, but there are also local objectives, e.g., to finish the job quickly and leave at five). Moreover, the top-down control that management attempts to wield will always be countered by other features within the organization: trade union pressure, the eternal battles for prestige and promotion among staff, and so on. There will always be internal political (in the broadest sense of the word) forces that will, even if only temporarily, overwhelm the major aim of the organization, even if this *is*, in fact, safety.

NAT also questions the major *technical* aspect of high reliability theory: redundancy. High reliability theory predicts that the more redundancy (i.e., duplication) of technical processes there are, the safer a system will be (because if one system fails, there will always be a backup). NAT, on the other hand, argues that the more redundancy you have in a system, the more difficult it is to understand and therefore the more likely that previously unforeseen, accident-causing interactions can occur between components. Finally, the presence of technical safety nets can lead to a false sense of security in which people take less care of what they are doing because they are aware of the safety features (this is one of the main features of risk homeostasis theory, pioneered by Wilde and Adams, who argued that people will be more careful if they know there are no safety systems than if they think there are such systems.. For example, will a tightrope walker be more or less careful if he or she knows there is no safety net, and that a slip might lead to death? [22]. A classic example was that of the heat sensor in the *Apollo 13* discussion.

Sagan's "revised" NAT also makes predictions regarding why a culture that learns will only solve some of the problems of safety. To begin, as we argued in Chapter 1, to attempt to localize the causes of accidents is to play an attributional game. In the politicized context of an organization, those who tend to have the most power will tend to control what attributions are made. Even when they do not, people can simply get things wrong and therefore learn the wrong lesson. Moreover, few, if any, organizations are completely transparent; there will always be coverups, misreporting, underreporting, and so on, all of which mitigate against organizations fully learning what needs to be learned.

Now, to avoid weighing the scales too far, we avoid simply listing all the accidents that have occurred in seemingly accident-proof systems. Needless to say, since the *Titanic* was proclaimed to be unsinkable, accidents and disasters have been occurring when experts proclaimed they were impossible. It is obvious that events like the *Titanic* disaster add support to NAT.

Instead, we ask a far harder question for NAT, one that would seem to argue against its acceptance: why haven't there been more nuclear weapons accidents? Given the nature of the system, NAT would seem to predict that accidents in this nonlinear, tightly coupled system would be inevitable. Why haven't there been?

Sagan's answer is simple: there have been no major nuclear weapons accidents simply because of blind luck. Again, this makes people in the secular West uneasy. Given the knowledge is power assumptions we looked at, the indeterminism we argued for in Chapter 1 is hard to accept. When something goes wrong (or, for that matter, when something goes right), we immediately look for the person or thing who was to blame (or the person to thank). The fact that some things might not really *have* a cause but might be simply "one of those things," the product of indeterministic, stochastic processes, directly contradicts much of what we were taught in school and college.

Sagan looked at a number of events to see whether an accident *might* have happened even if it did not in fact do so. The closer an accident came to happening, and the more random the actual safe outcome appears to have been with hindsight, the more evidence there would be for NAT. We now look at a specific example: the Cuban missile crisis.

6.4.6.2 Nuclear Accidents Avoided

To recap, the two competing theories have competing predictions for a specific situation. High reliability theorists would predict that there was little chance of an accidental nuclear war during the Cuban missile crisis, for five reasons:

1: There was an extremely strong commitment to safety from the head of the organization (in this case, President Kennedy).
2: The system was vastly redundant.
3: All the personnel involved were extremely highly trained and motivated.
4: There was a good safety culture.
5: High reliability theory (and, interestingly, risk homeostasis theorists) would predict that the chances of an accident would decrease precisely because the consequences of an accident were so huge. Everyone had an incentive to be safe because an accident might lead not only to someone else's death (which is frequently the case), but also the end of human life on Earth.

Normal accident theorists would predict, on the other hand, that there would be a high risk of accidents at this time for the following reasons:

1. The system was vastly complex. There were hundreds of thousands of military personnel distributed around the world. There were numerous control systems working at the same time, all of which had possibilities of interfering with one another. The system was deliberately opaque: due to the risk of espionage, the whole system was deliberately built so that one person could not gain the information needed to understand all of it.
2. The system was also tightly coupled. Because of the possibility of a preemptive nuclear strike that might "take out" the American military command, the system was designed such that retaliation to such a strike would be almost instantaneous. Of course, this was particularly the case at times of crisis.
3. Sagan's own version of the theory made some more predictions. The garbage can model predicted that, at times of high military alertness (at which *operational readiness* would take priority), safety might be pushed aside. Therefore, again we would see a higher risk of accidents than at normal times.

The Cuban missile crisis lasted for some time, and there is no space here to discuss the various possibilities for disaster that occurred. One will have to do.

At the time of the crisis, it was widely believed in the American military that before the Soviet Union attacked, it would send special *spetznaz* attack units to take out U.S. military bases that housed nuclear missiles. On the night of October 25, 1962, an air force guard at Duluth military base saw what appeared to be a Soviet soldier climbing the security fence round the base. He immediately pulled his gun and shot at the intruder, before running and setting off the alarm, which automatically triggered alarms at other bases. Immediately, armed American military personnel were sent to guard the perimeter fences of numerous air force bases in the area.

At Volk Field in Wisconsin, however, the alarm system was faulty. Instead of the intruder alarm, the alarm that indicated that America had been attacked by the Soviet Union was triggered. Because the United States was at that time on Defense Condition 3: Increase in force readiness above normal readiness (DEFCON 3), the pilots had been told that there would be no practice drills. They therefore had no reason not to think that they were now at war with the Soviet Union. The pilots of F106A interceptor craft were scrambled to protect the base. These were taxiing down the runway when a car drove onto the runway, lights flashing, forcing the aircraft to stop. An officer leapt out and explained to the pilots that Duluth had been in touch, confirming that they were not, in fact, at war with Russia. The retaliatory strike was therefore cancelled. The Russian saboteur turned out (rather ironically) to be a bear.

Of course, high reliability theorists could claim that because this did not, in fact, lead to a nuclear war, therefore the system was safe. So, the question is, how close did we come to catastrophe?

The key point here is that for a reasonably prolonged period of time, a military base was under the impression that nuclear war had begun. The reason this matters is that, at this time, various American bombers and fighters were sent to and from military bases (as one might expect). However, incredibly, the bases where they were to land were not kept up to date regarding where and when they were to turn up (this procedural oversight was altered as a direct result of the near-miss described here). Therefore, in conditions of fog or some other low-visibility situation, the first thing the base would have heard about an incoming American plane was when an unidentified aircraft appeared on the radar screen.

What normally happened in these situations, of course, was that air traffic control would quickly make radio contact, and the plane would be identified as American. The problems would have started in this specific situation if the interceptors had already been scrambled. Remember, it was only by chance that radio contact was made, sending out the car that stopped the planes on the runway. If this crucial call had been made just one or two minutes later, the planes would have been in the air.

So, now in the air, when the interceptors saw a mysterious and unknown plane looming out of the fog, they would almost certainly have thought it was a Soviet plane and, given that they thought they were at war, would have been well within their rights to shoot it out of the sky. This is not implausible. The year before, a B-52 had been shot down by accident by an F-100 interceptor, killing three crew members. Moreover, when security was at DEFCON 3, the interceptor had the right to use nuclear weapons. If this had occurred, given that the electromagnetic pulse would result in the loss of radio contact, personnel at Volk would almost certainly conclude that they were under nuclear attack from the Soviet Union. Moreover, anyone else who saw or heard the explosion would also assume that war had begun and take appropriate action.

This sequence of events (which, to point out the obvious, was unlikely but not impossible) would have almost certainly led to a huge loss of life, but it would probably not have led to nuclear war. Radar would quickly have noted that there were, in fact, no Soviet missiles entering American airspace. It is therefore highly unlikely that the president would have declared war.

The same cannot be said of the situation on October 28, 1962, the final (and tensest) Sunday of the crisis. At 9:00 A.M., the Moorestown radar base in the United States picked up what appeared to be a launching of a nuclear missile from Cuba. The computer quickly predicted a missile would hit Florida, with an estimated strike time of 9:02 A.M. The data were checked with Colorado (North American Aerospace Defense [NORAD] command center) and were confirmed: a missile was on its way. Because military personnel now had under a minute to react, they simply did nothing until after 9:02 A.M., when they could get in touch with Florida and confirm that no missile had struck. It transpired afterward that a test tape of a missile attack had been inserted into the Moorestown computer system (this was standard behavior to check the system) at precisely the same time Moorestown picked up a dummy signal from a satellite that *mimicked precisely the position and direction* of the test tape. Therefore, they became confused and mistook the dummy signal for the real signal and assumed that the test tape was now real. Before the error was found, Strategic Air Command in Omaha, NE, had been informed.

6.4.6.3 Do Bears Shoot in the Woods?

Taken on their own, these situations might not seem too frightening. After all, there was no war. However, taken in the context of literally dozens of such near-misses detailed by Sagan, they confirm, in many important respects, the predictions of NAT as opposed to high reliability theory.

In the first case, there were problems of nonlinearity and tight coupling. As we know from chaos theory, tiny causes can bring about large effects. This is a classic case. A bear moving toward a fence was interpreted as a fully armed Soviet *spetznaz* undercover agent (presumably the first of many) attempting to climb a fence and then neutralize a military base. In retrospect. of course, we see this as implausible; after all, we know what really happened. But, as we saw in Chapter 1, there are fundamental attributional biases that condition our behavior. We tend to assume (*ceteris paribus*) that if the effects are large, then the cause must also have been. So, it just does not occur to most people that the cause of scrambling nuclear armed interceptor jets was something as trivial as a bear. Once the situation is up and running, it is more plausible for us to assume that the cause must have been something significant (i.e., an invasion) rather than something trivial.

Moreover this is again a classic case of a safety net turning into a noose. In this case, the base was on such a high alert that almost anything could trigger the alarms. Moreover, because the system was tightly coupled, there was a fast reaction: planes were almost airborne before they could be stopped.

It also shows the problems of opaque systems. Personnel were horrified to find out afterward that bombers and fighters were being sent to land at bases without informing the base first. But, the system was so big and so complex that personnel at the bases simply had not been told. Moreover, because the system was huge and (deliberately) complex, no single person understood it all. Senior military personnel simply had not heard about this situation.

Again, in the second example, the safety system (against nuclear attack) was turned against itself again, we see the problems of *inherently unpredictable* systems. It had simply never occurred to anyone that radar might run a test tape at the precise time that it picked up a stray signal from a satellite. Moreover, the system was so tightly coupled that there was no time to do anything about it. If the situation had been different (if there had been some other malfunction that led personnel to think that they were already under attack), then Omaha would have had under a minute to decide what to do, not enough time to get in touch with the president (or anyone else). This situation might well have led to a massive nuclear retaliation against Cuba and possibly against the Soviet Union.

Despite these interpretations, it is still the case that there was no nuclear war (or at least has not been one so far, at the time of writing). Is it therefore possible to outline any further evidence for the predictions of NAT?

6.4.7 Normal Accidents in the Petroleum Industry

This final section of the chapter draws on work by Eli Berniker and Frederick Wolf, who have done important empirical work on NAT. Wolf and Berniker undertook a study of a five-year period (1992–1997) of 36 petroleum refineries located in the United States with a view to test NAT [see 25].

To repeat, Perrow claimed that systems that are tightly coupled and nonlinear will be particularly at risk of systems accidents. However, one of the failings of Perrow's work is that he never gave a technical definition of what he meant by *tightly coupled* or *nonlinear*. To a certain extent, therefore, these phrases remained "in the eye of the beholder" (given that these are taxonomic categories, of course this is exactly what one would expect). This was a useful approach for the discursive, sociological approach Perrow used in his book, but for empirical studies tighter definitions have had to be produced.

To begin, Wolf and Berniker made an assumption that builds on the theory of open versus closed systems; that is, the more complex a system was, the more likely it was to be un-understandable (open) by the staff in the system and therefore, to produce unpredictable (nonlinear) results. Therefore, Wolf and Berniker would have to come up with a technical definition of the word *complexity*. To do this, they divided up the processes performed by the refineries into 15 specific kinds of *unit process* (coking, asphalt production, alkylation, and so on) and defined the complexity of the system as the number of unit processes it contained. However, this was a crude measure and failed to take account of the fact that, like all open systems, refineries are dynamic, not static, systems.

Therefore, the unit processes were again subdivided into nodes, and parameters were defined that described the action of the system at that specific node. For example, one node might be concerned with the flow of a substance, and this could be categorized in a number of ways (e.g., no flow, partial flow, reverse flow, etc.). As such, an equation could be produced that took into account the amount of *energy* used by a specific node in the system, creating a variable C, defined as the number of all possible states of all parameters of all nodes for the plant. Therefore, the refineries could then have their complexity (and hence nonlinearity) measured and could be divided into two classes: less complex and more complex (i.e., less nonlinear and more nonlinear).

This left the problem of measuring whether coupling was loose or tight. Luckily, there are two specific methods of processing oil: continuous operation (pumping product directly into a pipeline) or batch (on and off) format (pumping product into tanks). It was hypothesized that refineries that used continuous processing would be more tightly coupled (because there would be much less slack in the system) than those using batch processing. Therefore, refineries could be subdivided again into tightly coupled and loosely coupled.

A measure of accidents was then needed. This was obtained from the U.S. Environmental Protection Agency's Emergency Resource Notification System Database. This measured all outages of hazardous substances in excess of reportable quantities (RQ). It should be noted that these RQ limits are very high. Therefore, anything reported in the database was by definition a very serious event (as a double check, data were also collected from two other relevant accident databases).

Initial data supported NAT. However, it was felt that the distinction between continuous and batch processing as a measure of tight or loose coupling was crude. So, Wolf and Berniker then made an assumption that has very interesting theoretical implications for NAT.

Following Shrivastava's research into the Bhopal disaster, it has been noted that organizations that receive less attention and resources tend to be less safe than those that *do* receive adequate resources and attention [23] This has been empirically tested by Nancy Rose, who, in a study of the American airline industry, found a correlation between lower profitability of airlines and higher accident and incident rates (especially for small carriers). More profitable firms had fewer incidents and accidents [24].

This suggests that there is, indeed, a safety-profit tradeoff. If the firm is making high profits, then money can be found to make the necessary improvements: order the latest technology, pay for regular servicing, pay for training, and so on. However, if profits are low, then these are precisely the features that will go by the board, and there will inevitably be more accidents

Wolf and Berniker hypothesized that firms with profits that were under pressure would be forced to tightly couple their operating processes. There would be less slack in the system. Time pressures would be more firmly emphasized; there would be more rigid adherence to deadlines. Moreover, there would be less money to buy the latest technology to facilitate the production process. Even when an accident did happen, the ability of the organization to learn from it would be hampered. Even if the accident investigation (for example) showed that component x would have to be bought to prevent further such accidents, there would be organizational pressure to delay such a purchase if it proved to be too expensive.

It proved possible to reanalyze the data and include more refineries. This time, the measure of coupling was profitability: the more profitable the company that owned the refinery was, the more loosely coupled it was measured to be. Again, empirical data were as predicted. Refineries that were more nonlinear, and more tightly coupled, had more accidents and incidents as measured on the relevant databases than those that did not. Furthermore, the kinds of accident were themselves more serious. Some normal accidents actually led to explosions and loss of life [25].

6.5 CONCLUSION

This has been a rather bleak chapter, but Wolf's amendments of NAT indicate at least a gleam of light in the darkness. To repeat, NAT demonstrates that accidents are inevitable and unpredictable. Nevertheless, because the accident rate cannot be reduced to zero, that does not mean that it cannot be reduced at all. Wolf noted that, during the period 1993–1997, three highly complex refineries reported no RQ events *at all*. All three were run by the same company. Of course, as with all statistics, one must interpret these results carefully, as Wolf pointed out. Is it that they have such a good safety culture that accidents do not happen or such a bad safety culture that accidents are "hushed up"? [26]. It is hard to say. Further research was in progress at the time of writing. Nevertheless, it does suggest that the deciding factor, not surprisingly, tends to be money. Perrow pointed out that even though the *Apollo 13* accident happened due to complexity, tight coupling, and the like, the reason it was recovered was because NASA had almost unlimited staff and resources it could throw at the problem, as well as state-of-the-art flight simulators that could try out recovery scenarios. This does suggest that resources and staffing play their part in making a system safe (massive resources did not help in the military nuclear defense scenario, but there you have another confound, the deliberate opacity of the system and the need for secrecy, which mitigated against organizational learning).

In any case, the statistical analyses of Wolf add more evidence that NAT is a worthwhile and useful tool of analysis, and as Wolf pointed out, the basic assumptions of the second law of thermodynamics (which were used to create his equation above) related directly back to von Bertalanffy's systems theory. Therefore, it seems that systems theory and cybernetics have some validation in the safety context.

The key point of this chapter has been to look at systems theory and its presupposition of emergence as a fundamental component of the sort of systems we tend to deal with in safety. Moreover, we have been keen to emphasize the rigor and empirical nature of the studies that test the theory. Many scientists get into a sort of existential panic when faced with the limitations of TR, assuming that the only alternative to TR is a sort of postmodern relativism or woolly minded mysticism. We are, clearly, arguing the opposite here. We are arguing that it is the technical rationalists who are fundamentally not living in the real world because methodologies, like everything else, are situation specific, and attempting to use methodologies taken from TR where they are not appropriate is quite simply using the wrong tools for the job.

There is a final point. If we remember the Confidential Incident Reporting and Analysis System (CIRAS) database (described in Chapter 2), then we will remember that it was a study of a large complex system (the U.K. railways) over time. The data set we worked with consisted of a hierarchical taxonomy that we treated as a *model of the system*. Analysis involved such tasks as looking at how these aspects interacted with each other; logging periods when the system was out of control; and examining emergent clusters where safety events seemed to have been underpinned by generic aspects of the system (see Section 4.5.2). What we are suggesting here is that taxonomy theory is a systems approach involving the creation of rigorous grounded models of complex systems that facilitate sophisticated statistical analyses without making any of the (false) presuppositions of null hypothesis testing, reductionism, Newtonian causality, and so on.

However, we must now go back to looking at methodology and examine how the new approach we have been looking at can lead to an analysis of a specific disaster scenario.

REFERENCES

1. Checkland, P., *Systems Thinking, Systems Practice,* Wiley, Chichester, U.K., 1996, p. 3.
2. Johnson, S., *Emergence,* Penguin, London, 2002.
3. Von Bertalanffy, L., *General Systems Theory,* Braziller, New York, 1976.
4. Wiener, N., *Cybernetics,* MIT Press, Cambridge, MA, 1965.
5. Cziko, G., *The Things We Do: Using the Lessons of Bernard and Darwin to Understand the What, How, and Why of Our Behavior,* MIT Press, Cambridge, MA, 2000, p. 60.
6. Cziko, G., *The Things We Do: Using the Lessons of Bernard and Darwin to Understand the What, How, and Why of Our Behavior,* MIT Press, Cambridge, MA, 2000, p. 61.

7. Lakoff, G. and Johnson, M., *Metaphors We Live By,* University of Chicago Press, Chicago, 2003.
8. Von Foerster, H., Cybernetics of cybernetics, in Midgley, G. (Ed.), *Systems Thinking,* Vol. 3, Sage, London, 2003.
9. Maturana, H., Reality: The search for objectivity or the quest for a compelling argument, *Irish Journal of Psychology,* 9(1), 25, 1988.
10. Luhmann, N., The autopoiesis of social systems, in Luhmann, N. (Ed.), *Essays on Self-Reference,* Columbia University Press, New York, 1990, pp. 21–79.
11. Waldrop, M., *Complexity,* Simon & Schuster, New York, 1993.
12. Dewey, J., The reflex arc concept in psychology, *Psychological Review,* 3, 1896, 357.
13. Kotchoubey, B., Event-related potentials, cognition, and behavior: A biological approach, *Brain and Behavioral Sciences,* submitted (available online at http:// www. uni-tuebingen.de/uni/tci/links/erp.pdf). See also Kotchoubey, B. and Lang, S., Event-related potentials in a semantic auditory oddball task in humans, *Neuroscience Letters,* 310 (2–3), 93, 2001.
14. Hollnagel, E., Nåbo, A., and Lau, I.V., A systemic model for driver-in control, in *Proceedings of the Second International Driving Symposium on Human Factors in Driver Assessment, Training and Vehicle Design,* Utah, July 21–24, 2003, p. 86.
15. Hollnagel, E., Cognition as control: A pragmatic approach to the modelling of joint cognitive systems, *IEEE Transactions on Systems, Man, and Cybernetics,* in press.
16. Flach, J., Beyond error: the language of coordination and stability, in Hancock, P.A. (Ed.), *Human Performance and Ergonomics,* Academic Press, London, 1999, p. 111.
17. Perrow, C., *Normal Accidents,* Princeton University Press, Princeton, NJ, 1999.
18. Perrow, C., *Normal Accidents,* Princeton University Press, Princeton, NJ, 1999, p. 90.
19. Perrow, C., *Normal Accidents,* Princeton University Press, Princeton, NJ, 1999, p. 96.
20. Sagan, S., *The Limits of Safety,* Princeton University Press, Princeton, NJ, 1993.
21. Cohen, M.D., March, J.G., and Olsen, J.P., A garbage can model of organizational choice, *Administrative Science Quarterly,* 17(1), 1, 1972.
22. Adams, J., *Risk,* Routledge, London, 1995; Wilde, G.J.S., *Target Risk*, PDE Publications, Toronto, 1994.
23. Shrivastava, P., *Bhopal: Anatomy of a Crisis,* 2nd ed., Paul Chapman Publishing, New York, 1992.
24. Rose, N., Profitability and product quality: Economic determinants of airline safety performance, *Journal of Political Economy,* 98, 5, 1, 944, 1990.
25. Wolf, F. and Berniker, E., Validating normal accident theory: Chemical accidents, fires, and explosions in petroleum refineries, *Journal of Management,* 16, 3, 1, 1990.
26. Berniker, E. and Wolf, F., Managing Complex Technical Systems Working on a Bridge of Uncertainty, paper presented at the International Association for the Management of Technology, Lausanne, Switzerland, March 2001.

7 *Challenger* and *Columbia*

We now look at two case studies, one in depth and the second a bit more briefly, in an attempt to try to provide some empirical evidence for some of the points made in the last few chapters. The second example will lead us into the next chapter, in which we attempt to make clear how the first half of the book (i.e., taxonomy theory) relates to the second half.

7.1 THE *CHALLENGER* DISASTER

One of the most famous (or notorious) disasters of our times was the space shuttle *Challenger* disaster, in which seven people (including a teacher, Christa McAuliffe, who was to have been the first "ordinary person" to travel in the shuttle) lost their lives. This has become famous not only because, in a media-saturated age, the event was caught on camera, but also because of what happened afterward. Everyone remembers that it was Richard Feynman who demonstrated that the famous o-ring tended to become brittle in cold weather, and that this was an important causal factor in the accident. Equally, everyone knows that it was discovered after the disaster that there had been a move to stop the launch that was overridden by senior management. This has become famous, therefore, as a very modern accident: a managerial accident.

7.1.1 SYSTEMS THINKING, SYSTEMS PRACTICE

Before we look at the *Challenger* disaster itself, we briefly recap the themes discussed in the last two chapters. In Chapter 5, we looked at the extent to which human beings existed in a systems relationship with their fellow human beings and the external environment, such that there were no abstract human beings, but people-in-a-car, people-in-a-plane, or whatever. Then, in Chapter 6 we looked at larger systems (or organizations) and saw the impact these could have on those events that, for sociocultural reasons, we call accidents.

Therefore, everyone works, acts, and talks within a specific context (this theory termed *contextualism*, we looked at briefly in Chapter 1), which is merely another word for system. To understand the meaning of what people do, we must therefore look at the system in which they are acting and of which they form a part.

We also looked (in Chapter 5) at the concept of indexicality as used in the sociological discipline of ethnomethodology. *Indexicality* is the idea that discourse and actions only take place within a certain context (system), and that when this context changes, the meaning also changes. So, for example, the word *bad* means, well, bad, but in certain street slang it also means good (when Michael Jackson named an album of his *Bad* he did not mean to imply that the album was terrible). Change the context, change the meaning; this is one of the backbones of taxonomy theory (Chapter 2).

There is another thing that follows from this that we have explained in two different ways: in the language of second-order cybernetics and the language of hermeneutics. Both of these attempted to make the same point, which is that when we try to understand something (discourse, actions, or anything else), this is not a passive action, but an active act of interpretation, and it will always be done from our own specific point of view (this philosophical view, remember, is termed *perspectivism*, also briefly discussed in Chapter 1; see also [1]).

These are the basic philosophical ideas that have been touched on in this book, and they are essential to understanding what happened in the *Challenger* disaster.

7.1.2 Objectivism and Safety

First, however, we have to clear a few preconceptions. As many philosophers and sociologists have shown us, the various views we take of the past, or rather the views we make of the past, are not the pure objective recitations of fact we sometimes take them to be. Instead, we sometimes find mythic ways of looking at things even in what seems to be objective historical discourse.

We have described the conventional view of the *Challenger* accident. It has mythic resonances. There was an accident, and then an expert (Feynman) discovered that "the powers that be" were corrupt or stupid. Therefore, truth was illuminated.

Now, it would be a mistake to imply that we are saying that because this story is mythic that that means that it is *false*. Many stories can be mythic and still true. But, what myths do is to simplify matters: make them assimilable to various cultural constructs. What the Feynman myth does is to play with some cultural assumptions that we look at in the next chapter, to do with experts and their function. So, it is not false.

But, it *is* an oversimplification. The National Aeronautic and Space Administration (NASA) was well aware that rubber becomes more brittle in colder weather. It was the more specific problem of what that fact *meant* in this particular *context* that mattered. To get a more rounded and broader picture of what happened, we have to use a methodology that cuts deeper than the ones normally used in a sociotechnical accident of this sort.

7.1.3 The Methodology

Let us start by looking at issues of methodology. As discussed in Chapter 1, the methodology you choose for looking at your subject matter will condition what you find. This is generally accepted, although the scale of the problem is not always understood. Now, for the discussion of *Challenger* we use as a source text *The Challenger Launch Decision* by the sociologist Diane Vaughan [see 4]. This is one of the most exhaustive studies of any accident of our time. However, because Vaughan is a sociologist, her work is rarely cited in the accident/safety management/accident investigation literature. This is a pity because not only does it contain insights that are normally missing from such texts, but also it has implications for how to carry out such investigations.

7.1.4 THICK DESCRIPTION

Vaughan at various points in her book states that the methodology she uses is that of thick description, a phrase borrowed from the anthropologist Clifford Geertz, who in turn took it from the philosopher Gilbert Ryle. So, what does *thick description* mean? Geertz explained:

> Consider, he [i.e., Ryle] says, two boys rapidly contracting the eyelids of their right eyes. In one this is an involuntary twitch, in another, a conspiratorial signal to a friend. The two movements are, as movements, identical: from an "I-am-a-camera" "phenomenalistic" observation of them alone, one could not tell which was twitch and which was wink, or indeed whether both or either was twitch or wink. Yet the difference, however unphotographable from a twitch and a wink is vast, as anyone who has had the misfortune to mistake one for the other will know. [2]

Geertz suggested the presence of a third person who *parodies* the involuntary twitch boy for the purpose of ridicule. We now, therefore, have three different boys carrying out precisely the same action in precisely the same way. Yet, the meaning of this action is completely different in each case, and it is different because the context is different in each case.

By *thin description*, therefore, Geertz means simply a neutral description of the *facts* of the matter, without any context; in this case, the boy lowered an eyelid. This tells us the facts but no more. *Thick description*, on the other hand, provides the *context* as well so that we can make sense of the action, so that we understand what it *means*. The links between this view and contextualism and indexicality should be obvious. The object we perceive is always in a context. Equally, the perceiver is also in a context. Both these contexts will influence how an event is interpreted (i.e., taxonomically classified).

7.1.5 SIMPLIFICATION, DIAGRAMS, AND MAPS

A thick description of something like the *Challenger* disaster will, therefore, include the cultural context of what happened, and would make the point that *Challenger* was not some free floating accident. It was, instead, a specific incident that occurred in a specific situation, with specific people. In order to understand the event we have to see what these people were doing and saying before the accident, without falling victim to what is termed in social research as the fallacy of retrospective determinism, the idea that things had to be that way. This is closely allied to hindsight bias: if the *Challenger* accident had to happen, how could people have been so stupid not to see it coming? The fact is that the accident did *not* have to happen, and that for everybody at NASA (and elsewhere) the launch of the *Challenger* was simply another day at the office. We invest the events leading up to the accident with deep significance because we are looking at them from a certain historical perspective.

We choose to interpret certain events as causal factors, precursors, or whatever. But, these were not a focus for anybody at the time.

If the reader's working life is anything like ours, it will consist mainly of coffee breaks, hunting for the stapler, talking on the phone to technical support because the computer has a virus again, breaking for lunch, and so on. In other words, "just one damn thing after another." This should make us aware of another fallacy, the fallacy of misplaced concreteness (sometimes also known as the Mayan illusion*), which the Polish linguist Alfred Korzybski succinctly described as "mistaking the map for the territory." In numerous books on *Challenger* (and other disasters), we are presented with a series of neat boxes with arrows pointing between them, indicating that there were discrete causal factors that neatly interacted to cause the accident. Now, we may use these diagrams to clarify our thinking if we wish, but it would be a huge mistake to delude ourselves that *that is what actually happened*, and that life *really is* neat little causal factors interacting. Real life is a good deal messier than that, and what can be seen and interpreted depends on where you are standing. For better or for worse, we all stand *after* the *Challenger* crash, and we interpret the events leading up to it in that light, that is, as causal factors leading inexorably to an accident. We must always remember that, to repeat, *that is not how it looked at the time*.

7.1.6 THE DISASTER ITSELF

The basic facts of the disaster are not in question. The space shuttle *Challenger* was launched on January 28, 2004, at 11:38 A.M. The spaceship disintegrated 73 seconds later. All seven crewmembers died.

To fill in some other important background, as most people know, the shuttle is launched with a solid rocket booster (SRB) to propel the shuttle out of Earth's atmosphere. In the 1950s and 1960s, it had been accepted that all material to do with the space program would be kept in-house, that is, made by NASA itself. However, from the early 1970s onward, pressure from market forces began to build to outsource various tasks to external contractors to cut costs. The manufacture of the SRBs was one of these outsourced tasks. In fact, four firms had bid for the contract to make the SRBs, but the firm that won the contract was Thiokol-Wasatch, a subdivision of Morton-Thiokol. Although Thiokol (as we refer to the company throughout) was not a market leader, it had had a reasonable amount of experience in the field, manufacturing Minutemen and Trident missiles for the U.S. military.

However, in the accident investigation held after the disaster, it quickly became apparent that it was a flaw in the SRB that had caused the accident. Specifically, it was a joint or seal called the o-ring in the solid rocket motor (SRM) of the SRB. Made of a black rubberlike substance called Viton, the o-ring is about a quarter inch in diameter. The seat in which it is placed is lined with asbestos-filled putty when in the SRM to protect it from the high temperatures and exhaust blast it will sustain during the initial stages of flight. The o-rings are needed because the SRB is manufactured in units that are then assembled by Thiokol and shipped to NASA. Under normal circumstances, there are no gaps between the various units when they

---

* *Maya* is the Sanskrit word that means the error of mistaking the symbol for the reality.

are put together, but under the high pressures of ignition, a tiny gap develops and is sealed by the o-ring (to be more specific, if all goes well, a tiny gap is not allowed to develop because of the o-ring). The o-ring, therefore, keeps the (burning hot) propellant gases inside the SRB.

However, what quickly became apparent is that one of the o-rings of the right SRB had not done its job. Ignition gases got through the o-ring and eventually set fire to the hydrogen fuel of the external fuel tank. This caused the disaster.

When this information became publicly known, it was bad enough, but things quickly became much worse for Thiokol (and NASA). It turned out that, on the night before the launch, there had been a presentation by various members of Thiokol in which they had expressed their concern about launching the shuttle in low temperatures. They were particularly worried about the effect this might have on the o-ring. As Richard Feynman pointed out later, the whole point of making the o-ring of a rubberlike substance is that it would retain its elasticity to plug the gap between the units in the rocket. In cold weather, however, things become brittle. Perhaps, it was suggested at this prelaunch meeting, the o-ring would become too brittle to plug the gap and thereby contain the burning hot propellant gases. Perhaps the gases might get through. This, of course, is precisely what happened.

The subsequent investigation discovered four fundamental problems with the o-rings. First, there was the tendency for blow holes to develop in the rings, enabling exhaust fumes to penetrate between the rings and the cylinder. Second, there was the fact that the rings tended to erode under the constant pressure of the propellant gases. Third, there was something termed *join rotation*, which stretched the gap the o-ring had to fill. Finally, and crucially, there was the fact that the dynamics of the o-ring were temperature specific. In other words, the rings had less and less elasticity and give the colder it got.

However, the real nail in the coffin for Thiokol's reputation was that it quickly became clear that all these features had been known (or suspected) before the crash. How, then, had the launch been allowed to proceed?

7.1.7 THE CULTURAL BACKGROUND

No event exists in itself; it always exists in a cultural and historical context. To explain further what happened, we have to look at the background of the *Challenger* disaster.

The space shuttle was born out of the confusion over NASA's role in the 1970s. Whereas in the 1950s and 1960s NASA had been given, essentially, unlimited personnel and resources, in the 1970s the war in Vietnam and the increasingly skeptical public mood meant that a new attitude began to develop among politicians. Increasingly, it began to be felt that NASA would have to pay its way, and that cost-effectiveness would be an increasingly important aspect of any further missions.

The space shuttle was ideal in terms of this new philosophy. Not only was the fact that it was reusable and would lead to greater efficiency, but also it was argued that the shuttle would fit neatly into military plans for the militarization of space, especially in terms of reconnaissance satellites. When the shuttle was designed, therefore, it was built with military usage in mind.

However, almost immediately, problems were encountered. NASA wanted a relatively small spacecraft (suitable for civilian payloads); the military wanted a large one (for military payloads). When NASA had begun to plan the shuttle, its original plan had called for two piloted winged vehicles that would be bolted together at launch. After launch, the big carrier would release the shuttle proper, which would continue up into space. Both craft would be fully reusable.

However, the military's insistence that the shuttle should be able to carry not only any military equipment in use at the time, but also any equipment that might be built at any point in the future ruled this out. It also meant that the shuttle would have to be *much* larger than NASA had anticipated.

Not only this, but the Air Force also insisted (following the flight style of its previous spacecraft [e.g., the X-20]) that the shuttle should be able to maneuver extensively during reentry. Due to the design features that this would entail, an orbiter designed to suit the Air Force would not only need to endure far greater structural and thermal loads (i.e., than NASA had planned) during reentry, but also would make an exceptionally steep descent, and it would touch down at a correspondingly higher speed [3].

By the 1980s, therefore, the whole shape, purpose, and flying ability of the shuttle had now been changed by financial and political pressure. Moreover, the ditching of the fully reusable plan (which would have meant the use, in effect, of two shuttles) meant that NASA was now committed to the use of an SRB that would be attached to the shuttle's external fuel tank. Even though NASA did not want this design, it had a major factor in its favor, which was that use of the SRBs would cut down costs (as an unmanned rocket would obviously be cheaper to run than a manned [and built from scratch] shuttle booster). The decision to use a *solid fuel* booster rocket was also cost driven.

One consequence of all this compromise was crucial: all previous manned American spacecraft had carried small escape rockets that the crew could use in the case of an emergency at the start of a mission. Horse-trading over the size, cost, and thrust of the SRBs meant that the escape rockets had to be ditched to save weight.

The end result of all these compromises was a shuttle that nobody really wanted, but that everyone now felt they had to defend.

7.1.8 THIOKOL AND THE SRB JOINTS

Before discussing the question of how the o-rings were actually designed, we should interpret the history of the development of the shuttle in terms of systems theory. As we have seen before, everyone and everything stands in a systems relationship, in theory, to everything else. As the saying has it, everything is interconnected. For the purposes of analysis, we have to cut down the amount of data we have available and look at smaller systems. However, we should not be under any illusions. In the real world, systems interact, and system borders are fuzzy. Therefore, the worlds of economics, politics, science, and so forth are not autonomous. In reality, every scientific decision is *also* a political and (therefore) *also* an economic decision (and vice versa). Hence, the futility of trying to ignore or disparage political and economic issues in terms of accident causation leaves one with an impoverished analysis.

Moreover, given that *Challenger* was by definition a systems accident, we can ask a further question: was it also, therefore, a normal accident in Perrow's use of that phrase? If it was, we should be looking for two specific things: nonlinearity and a tightly coupled system. In terms of nonlinearity, we should look for opaque systems (i.e., systems that were open; un-understandable, although they might be mistaken for closed and therefore understood systems) and failures due to redundancy. Recall that failsafes in normal accidents function as accident causers because they lead to a false sense of security that actually encourages accidents and disasters. Keep these concepts in mind as we look deeper into the causes of the accident.

To return to the o-rings: the SRB contracts were given out to Thiokol, and when it began to design the SRBs, it looked back and searched for a design it could copy. As we discuss in the next chapter, this is what experts do: they look back for previous situations and try to adapt what worked then. In this case, Thiokol decided to use the Titan 3 rocket as their model, with good reason. The Titan 3 is one of the most reliable rockets ever produced. Now, due to their size, the SRBs would have to be built in separate units and put together at NASA. Therefore, there would be a potential gap where the units would be joined, and it was this gap that the o-ring of the Titan (and, now the SRB) was designed to plug. However, when designing the SRB, Thiokol decided to improve the design of the Titan to make it even safer to cope with the higher risks of space flight. The Titan had one o-ring, but Thiokol's SRB was to have two: each primary o-ring for main use was to have a secondary o-ring as a backup in case the first o-ring failed.

However, in adapting the design of the Titan for use in the SRB, Thiokol ended up making a number of changes. One of these, ironically, was because of the addition of the second o-ring. The bottom rim of the fuel unit (termed a *tang*) had to be elongated to fit the second o-ring. This had a knock-on effect on the rest of the design. For example, the fit between the various segments of the SRB was no longer as tight as in the Titan, so gaps had to be filled with putty.

However, the main difference was the entirely different use of the Titan o-ring and the Shuttle o-ring. *The Titan was not a reusable rocket: it was used only once. The Shuttle's SRBs were designed to be used up to 20 times.*

The key point was that, although this new o-ring looked like merely an adaptation of the Titan design, and although the engineers involved regarded it as such, it was not. It was a completely new design to be used in a completely different situation (i.e., a situation for which the key was reusability). The o-ring seals had moved for all intents and purposes from a closed system (in which everything, more or less, was known about it) to an open system, which might function in unexpected ways in unusual situations.

7.1.9 Joint Performance

The first sign that this might be the case occurred during research in the early 1970s; engineers calculated that at ignition, the SRB joint might momentarily rotate, and that this might create a gap between the tang and the clevis (the two sides of the joint that the o-ring helped to seal). Now, this was not *necessarily* a problem, but experiments were nevertheless carried out to see whether there might be a risk.

In September 1977, the prototype joint mechanism was sealed, and it was indeed discovered that under pressure of ignition, the tang and clevis separated briefly (because of the rotation), creating a gap. But, what did this mean?

Here, we have to return to the concepts of the social nature of science we looked at in Chapter 1 and what has become known, in an ungainly fashion, as the social creation of a scientific fact. As we know from the hermeneutic tradition, everything, including scientific experiments, is not passively transmitted into the brain, but instead is actively interpreted by flesh-and-blood human beings in a social context. So, whereas (in terms of these experiments) there was no argument about the facts, there was a difference of interpretation about *what the facts meant.*

The problem here was a problem that we looked at briefly in Chapter 1: ecological validity. Experiments are carried out, like everything else, in a specific context. How do we know, then, that when the context changes the results of the experiment will stay the same?

Specifically, the issue was that in the experiments the o-ring was "put through its paces" 20 times, and 8 times in a row (at the beginning of the experiment) the o-ring functioned perfectly. It was only after it had been weakened by the previous tests that the o-ring began to fail. But, Thiokol argued that in reality the o-ring would only face ignition pressure once (i.e., when the Shuttle took off), not 20 times. Moreover there were other reasons why Thiokol doubted the ecological validity of the test.*

The experiments continued, but then another problem reared its head: conflicting results. Again, this phenomenon is well known by scientists and has been revealed by sociologists of science, but it tends not to turn up in official histories of how science ought to work. The debate here was about the size of the gap. Obviously, the larger the gap was, the more exhaust fumes that could get past it and the more dangerous the situation was. The problem was that Thiokol's experiments tended to show a small gap, while the experiments carried out by other engineers tended to show a larger gap.

However, the deciding fact was the presence of the secondary o-ring. Thiokol reasoned that, even though there might be a brief gap caused by joint rotation, *this did not really matter because of the second, failsafe o-ring.* Even if the first o-ring failed, they argued, the second o-ring was bound to work and stop any potential disaster.

We begin to see the origins of a classic kind of normal accident as discussed by Perrow. Not only did the second, failsafe o-ring cause structural changes that increased uncertainty (and therefore the possibility of danger), but also *the very fact that it existed led to a false sense of security* among the (Thiokol) engineers. Naturally, engineers outside Thiokol had differing views, but they could never penetrate this illusion of safety. Of course, in books and in writings by experts who write about what science is (or is meant to be), *experiments* are meant to sort out these differences of interpretation. In reality, they do not. Experiments can give conflicting results. Experiments suffer from problems of ecological validity.

---

* For example, the experiment was done in a laboratory (i.e., horizontally), whereas on the launch pad the joint would be vertical. Thiokol argued that this would take pressure off the joint.

Experiments have to be interpreted, and Thiokol consistently interpreted experiments as less significant (i.e., in terms of safety) because, to repeat, it was convinced that if there ever was a problem, it could not be a *serious* problem because of the failsafe, second o-ring.

There was another problem. NASA (and Thiokol) is an organization, and organizations are hierarchical. There are explicit social rules regarding who is allowed to speak, and there are implicit social rules regarding who will be listened to. The main engineer pointing out difficulties with the SRB joint, Leon Ray, was, in his own words, "way, way down in the organization" [4]. This question of who is allowed to speak, and who will be listened to, will become increasingly important as the story develops.

Eventually, however, Ray managed to get his boss, John Miller, to write to Thiokol expressing his concerns about the joint separation and some other problems Ray had with the joint. Ray recommended a redesign of the entire joint as a long-term solution to the problem. However, he also suggested numerous short-term fixes that were eventually carried out.

Nevertheless, Ray continued to test the joint (that is, the joint *before* it had the short-term fixes) and eventually carried out some experiments that suggested, for the first time, that under certain circumstances, *both* o-rings might fail (important note: by *fail*, we mean deviate from optimum performance level in a significant way; Ray did not prove, or claim to have proved, that therefore the space shuttle might actually blow up). However, again, Thiokol questioned the experiments, suggesting that Ray must have made a mistake. Ray denied this, but then ran into another problem. In normal engineering work, the work is heavily rule bound (something we look at in the next chapter). Normally, if one wishes to know whether the amount of joint rotation is acceptable in an o-ring, then one simply looks up the appropriate book and checks. If it is over the limit, then the design is changed.

But, because these were now innovative designs, *there were no rules*. Even assuming that Ray had shown that o-ring performance violated the standards for the Titan 3, this said *nothing* about whether it would meet the requirements for the shuttle. Eventually, Ray (or rather his company, Marshall) and Thiokol had to get together and create a compromise new rule that would henceforth define acceptable performance. After extensive testing (together this time), both Ray and Thiokol were satisfied that, with the changes that had been made, the o-rings now met the new standards or rules.

The key point, we must remember, is that these alterations had been recommended as stop-gap measures. No one at Thiokol responded to Ray's request for a fundamental redesign of the joint.

Moreover, as we have seen, over and over in this book, the activity of everything, human and machine, is context specific. At the end of the day, Ray and Thiokol interpreted the experiments in two very different ways. Ray interpreted them in a highly context-specific manner; that is, he argued that this demonstrated that the o-rings would not fail *in the specific circumstances tested*. Thiokol, on the other hand, interpreted this as indicating a general, non-context-specific law. It argued that the experiments proved that the o-ring seals *would function under any (normal) circumstances*.

This is due to another phenomenon known from the philosophy of science: overdetermination. It is well known that in theory all scientific experiments are overdetermined, but this becomes more apparent in the human sciences. *Overdetermination* is "when there are two or more sufficient and distinct causes for the same effect" [5]. An example from the same source might make it clearer: "A convicted spy stands before a firing squad. Two shooters have live ammunition, and each, at precisely the same time, succeeds in shooting the spy in the heart. Either bullet on its own would have killed the spy in roughly the same manner. The spy dies." In other words there are two theories here: theory A (that Shooter 1 killed the spy) and theory B (that Shooter 2 killed the spy). How do we decide between them?

The answer is, frequently, we cannot. They are both true. Therefore, we can have experimental results that have more than one possible explanatory theory. This should remind us of the possible attributional paths we discussed in Chapter 1, especially Lewontin's example of deaths by cholera. Lewontin, remember, demonstrated that there are two possible causal paths for the fact that cholera fatality rates declined throughout the 19th and 20th centuries, the sociological and the medical explanations, and that there are no criteria to determine between the two: they are both equally true. In other words, the decline in cholera fatalities is overdetermined.

In a similar way, the experiments discussed here were also overdetermined. Marshall argued that they showed that the o-ring was not safe; Thiokol argued the opposite. More experiments could be done (and were, as we shall see), but they did not settle the matter. They merely led to more confusion as the experiments were inherently polysemous. They were interpreted by different people to mean different things at different times, and no social consensus emerged regarding what they really meant, if anything.

7.1.10 RISK

The situation gets worse. We argued in Chapter 4 that risk is not an objective phenomenon, but is instead socially structured. *Challenger* is a classic example of that. Thiokol's social construction of risk was that, because of redundancy (the failsafe of the second o-ring), the SRB joint was now for all intents and purposes safe. It argued this because of certain assumptions it had made and because of the social context in which it was operating. Even though others challenged this cultural construction, it proved flexible enough to bend but not break: Thiokol did eventually admit that there was a problem but thought that the alterations it had made at Ray's insistence had solved the problem.

The rest of the story of *Challenger* is of this cultural construction bending even further as more and more anomalous data were adapted to this narrative, and data that seemed to argue against it were explained away or ignored.

7.1.11 FLIGHTS, RULES, AND WAIVERS

The first flight of the shuttle took place on April 2, 1981. After the launch, the o-rings were inspected. There was no evidence of leakage. It seemed that Thiokol's view had been correct. The joint was safe.

However, on the second flight, Thiokol engineers discovered that hot exhaust gases had eroded some of the primary o-ring. This was completely unexpected (by Thiokol). They discovered that the problem was again caused by the addition of the second o-ring. Because this had changed the dynamics of the seal, putty had been inserted to help seal the gap and protect the o-rings from direct exposure to exhaust fumes. However, it seemed that tiny bubbles had formed in the putty, which left weak spots. The exhaust fumes had gone through these weak spots, which had actually concentrated the heat on specific spots on the o-ring, partially burning through them. However, erosion occurred on only 1 of the 16 primary o-rings, so the problem was thought to be minor. The constitution of the putty was altered to prevent blowholes, and the problem was considered solved. Again, engineers stated explicitly that even if the first o-ring failed, the second o-ring was a failsafe.

It would be nice to believe that Thiokol in some way hushed up problems with the o-ring (and that this, therefore, was a problem of corporate malpractice), but it did not. In fact, it was known to be an issue, and when there was a routine change of staff at NASA, the new manager, Lawrence Mulloy, insisted that the joint seal no longer be considered fully redundant (i.e., failsafe). However, again experiments were ambiguous, and this interpretation depended on the point in the launch at which the exhaust fumes reached the o-ring. Moreover, Thiokol argued that, even if there was a risk, it was a very small risk, and that this eventuality was very unlikely.

This argument, of course, came under heavy criticism after the disaster, but again it only illustrates that engineers live in the real world. As John Adams demonstrated in his classic work *Risk* [6], literally everything involves risk. If you go out, then there is a risk; equally, if you stay in your house, then there is risk. If you drive, then there is risk, but if you choose to walk, cycle, fly, or stay in the same place, then there is also risk. Life *is* risk. The question is, what risks are you prepared to take? To say that flying the space shuttle is risky is to say nothing. Of course it is risky. The question is, to what extent are these risks acceptable? Here again, we come down to the fact that risk is, to a certain extent, a cultural construction. What strikes me as an acceptable risk will not necessarily strike you in the same way. Risks are perceived differently across cultures as well. Thiokol interpreted the risks as reasonable and acceptable because it had a certain view of the matter and because it had a certain history with the project. No malice was involved.

Soon after he changed jobs, Mulloy changed the status of the joint seal from a C1R (redundant) to a C1 (not proven to be redundant), but then he issued a waiver that permitted the shuttle to fly. After the disaster, this seemed to be a clear example of a *rule violation*. We investigate the roles or rules in the next chapter, but suffice to say at the moment that if the investigators had discovered a clear rule violation, they would, they felt, have come much closer to discovering the true root cause of the accident.

However, here we must remember indexicality. To put it bluntly, the word *waiver* within the culture of NASA does not mean what it means outside it ("change the context, change the meaning"). The accident investigators (and the general public) never fully understood this and persisted in acting as though a rule had been broken.

All that C1 meant was that a component had no *failsafe*, but clearly not every component *could* have a failsafe. The right wing of the shuttle, for example, did not have a failsafe. If it dropped off, there was not another right wing above it to take over. So again, things had to be interpreted. What were the risks of flying if this particular component did not have a failsafe? We should remember that at no point had an actual o-ring failed during flight. Except for (disputed and ambiguous) experimental results, there was no reason to think that any o-ring *would* fail. A waiver was a technical procedure to keep the shuttle flying while at the same time ensuring that the component in question received adequate attention. A waiver was a technical procedure that enabled a component to be passed for flight even though its redundancy could not be proven. Mulloy did not waive a rule (i.e., in ordinary language, ignore or override it). He used a waiver, a noun that, in "NASA-speak" *was* a kind of rule.

The flights continued.

7.1.11.1   Flight 6

On April 4, Flight 6 took place. Afterward, engineers discovered that again heat (exhaust fumes) had reached the o-rings (it must be stressed that at this point, just because heat had reached the o-rings, this does not mean that they were eroded, let alone penetrated or destroyed). Again, the putty was changed, and for the next few flights, there were no problems.

7.1.11.2   Flight 10

However, on Flight 10 (STS 41-B) *two* primary o-rings were not only heated, but also eroded. This was the first time that two joints had experienced erosion. Again, the engineers decided the putty was to blame. Again, it must be stressed, even the fact that there was erosion did not mean that the joint failed to seal or anything even close to that. The Thiokol engineers were observing that a new seal was behaving unusually, but paradoxically this was, in itself, not unusual. The joint was new: no one really knew how it was going to behave. With nothing for comparison, the engineers decided that this must be normal behavior for this kind of joint/seal. Certainly, again, as long as the failsafe worked, there did not seem to be a problem. However, this changed with Flight STS 41-D on the August 30, 1984.

7.1.11.3   Flight STS 41-D

The flight on August 30 was significant because it was the first time that blowby (soot behind a primary o-ring) was observed. This was highly significant because it suggested that, briefly, the exhaust fumes had actually penetrated beyond the first o-ring. Why, therefore, did the engineers not demand a halt to further flights until the problem was fixed?

The answer again was *because* of the failsafe. Thiokol engineers had now decided that the joint was only to be designated as risky if it looked like exhaust fumes might penetrate the *second* o-ring. Now, the engineers' view of the joint had altered. They now considered it *normal* if there was some form of erosion. Events that had caused alarm earlier (such as exhaust fumes reaching the seals) were no longer even considered remarkable. Slowly, the concept of safe drifted away from the original definition. *Safe* had originally meant no exhaust fumes reaching the o-ring. Now, exhaust fumes, erosion, even penetration of a primary o-ring *by* exhaust fumes were considered unremarkable. As long as the joint was still considered failsafe by Thiokol (of course, some doubted even this, but they were outside the Thiokol corporate structure and hence outside the Thiokol culture), then the shuttle would be allowed to fly.

7.1.11.4 The Culture of Drift

So, things proceeded. However, on the first launch of 1985, STS 51-C, initial blowby of the *second* o-ring was seen. For the first time, redundancy was challenged. (Significantly, this was at a time of extreme cold, but this was not remarked on at the time.) Again, why was the shuttle allowed to fly?

The answer again was because of the core belief of redundancy. Each time the shuttle flew and did not blow up, it gave credence to the belief that it would never blow up. Each time it flew, it seemed safer. In the same way, each time the o-rings took punishment but did not disintegrate, the sturdier they seemed. Whereas we, as outsiders after the fact, see the increasing damage the o-rings were sustaining, the Thiokol engineers saw the fact that they could take this damage and still function as a sign *that they were a good design.*

The salient fact was that the exhaust fumes would need a certain amount of time to burn through both seals. So, the question shifted again. The issue was no longer whether the o-rings would seal. It now became whether the two o-rings could seal *in time*.

The answer to this question was "yes," or so it seemed. The first o-ring had sealed before the second one had become eroded, which seemed to demonstrate that, with two seals, the exhaust fumes *would not have enough time* to penetrate through both seals.

Moreover, the extent of the erosion in each individual seal was less than it had been on other flights. If the erosion had been acceptable then, why not now?

There was also the effect of the cold. Contrary to what was implied by Richard Feynman and others after the fact, everyone at NASA knew that rubber and putty become harder when cold. However, the fact that the Florida temperatures at the time were unusually cold (indeed, unprecedented) again persuaded the engineers that the risk was low. If the temperatures were unusually low, then it was unlikely they would happen again. Therefore, the flights should continue.

Everything seemed to be fine until mission STS 51-B on April 29. On this mission, after the fact, the engineers discovered that, for the first time, the first o-ring had burned all the way through, and that the second ring had sustained severe erosion damage. This was immediately perceived as extremely dangerous, and a launch constraint was announced, which did not stop all flights but triggered an automatic investigation.

However, this investigation discovered a specific cause. It was discovered that the primary seal had been installed with some foreign matter (possibly a human hair) that had prevented it from ever sealing properly. Naturally, the engineers decided that this was the cause of the event, and that as long as precautions were taken to prevent this happening again, the o-ring was safe (again a classic example of overdetermination, as discussed in section 7.1.9, or polycausality; Thiokol treated this event as if it was monocausal and that removing the hair would solve the problem).

7.1.12 THE FINAL MEETINGS

We now move to the final meeting, which has since become notorious. However, we must remember that what seems clear to us (that the shuttle was in serious danger) did not seem clear to everybody at the time. What we should be looking at instead are the group dynamics of the meeting: who got to speak, who did not, and how the social construction of the risks involved was created.

The shuttle was originally due to be launched again on January 22, 1986, but the launch date was postponed a number of times for technical reasons. Eventually, it was decided to launch January 27 at 9:38 A.M. Stan Reinartz, the manager of the Marshall Shuttle Project Office, called relevant personnel to ask whether they approved the new flight time (this was a standard procedure). Everybody did, but Marshall's SRM manager, Larry Wear, had a vague recollection that low temperatures had preceded the flight of STS 51-C in January 1985, that Thiokol had made some remarks about low temperatures, and so asked for some more time to make inquiries. He asked the Thiokol Space Booster Project Manager Boyd Brinton whether he remembered anything about low temperatures, but Brinton could not. Wear asked Brinton to call Thiokol in Utah to see whether they had any background data about temperature and the performance of the shuttle.

As a result of this call, Thiokol's Bob Ebeling called a meeting in Utah and asked the relevant engineers what they felt about the effect of cold on shuttle performance. A few mentioned that tests had shown that the o-rings become more brittle in low temperatures, and that this might have safety repercussions. However, it was stressed that any safety concerns would depend very strongly on what the temperature actually was when the shuttle was launched.

After finding out that the temperature was forecast at 26°F, Ebeling decided that another meeting would have to be called, this time with NASA involved, to discuss the possibility of this particular temperature having an effect on the o-ring.

Actually, Ebeling arranged two meetings. The first was a teleconference at 5:45 P.M. that evening (at NASA). By all accounts, this was an unstructured affair in which Thiokol representatives merely expressed an opinion that cold might have an impact on the resilience of the o-ring. However, the line was bad, and it was decided to have a further meeting at 8:15 P.M. that night with additional personnel there to discuss the matter further. It was also decided that Thiokol would make a more formal presentation to put forward its concerns.

By this time, of course, the problem was that many of the relevant personnel had gone home for the day. However, eventually enough were found for the conference to go ahead. Meanwhile, Thiokol began to write its presentation.

After a prolonged discussion based, it should be stressed, on uncertainty, Thiokol eventually decided that it would not recommend a launch unless the o-ring temperature was equal to or greater than 53°F, which was the lowest temperature in their flight experience base. The reason this temperature was picked was because the flight that had flown at this temperature (STS 51-C) had experienced the worst blowby. Thiokol quickly prepared 13 charts for its presentation; such was the shortage of time that they were handwritten.

However, looking through the charts immediately before the teleconference was due to begin, a few people noticed something that had not been obvious from the individual charts: there were two known cases of blowby, and it was true that one of them happened at 53°F, but the other one occurred at 75°F. The engineers decided that the blowby at 53°F had been substantially more severe than that at 75°F; at the same time, they were aware that this weakened their case.

7.1.12.1 The Teleconference

The teleconference began 20 minutes late due to the time constraints on Thiokol. However, it eventually argued that experimental data confirmed that the colder the temperature became, the more brittle the o-ring would become. Therefore, it was no longer convinced by redundancy: it felt that there was now a chance that the second o-ring would not seal, and the flight should not go ahead unless the temperature was 53°F or above.

Now, this was the first time in the history of the shuttle program that Thiokol had come forward with a no-fly recommendation. However, it should be stressed that *it was not asking for the flight to be cancelled.* Instead, it was merely asking that it be postponed until a warmer day or even launched in the afternoon instead of the morning. In theory, none of the Marshall engineers/managers had a problem with this. The problem they had was that Thiokol's case was weak. It was not only weak in the sense that it looked unprofessional (handwritten instead of typed up), which should not have mattered but which unfortunately did. It was also weak in the sense that, as Thiokol admitted, the blowby was not caused by temperature in a simple deterministic way (i.e., the sort of Newtonian way that engineers like). Instead, it was one factor among others. Sometimes, low temperature seemed to cause blowby; sometimes it did not.

However, the key point was the arbitrary 53°F threshold. To Marshall, this number seemed to have been plucked out of the air. No one had suggested a low-temperature rule regarding the o-ring before. Moreover, it contradicted other available data. The manufacturer's guidelines for the material out of which the o-ring was made (Viton) suggested that it should be safe above 25°F. Moreover, even though previous shuttle launches had been at temperatures below 53°F, Thiokol had never even raised this as a possible issue, let alone suggested cancellation or postponement of the mission.

The debate went on for some time, until George Hardy (of Marshall) was asked for his view. He replied, in words that became infamous after the disaster: "I will not agree to launch against contractor's recommendations."

There was, apparently a dead silence after this, which perhaps is indicative of the fact that until he said these words nobody in Marshall had seriously thought that the flight might be postponed. Then, Thiokol asked for five minutes offline to discuss the matter further.

This discussion ended up lasting 30 minutes, however. It should be stressed that at this point Marshall engineers and managers, even though they were surprised at the way the discussion had gone, were perfectly prepared not to launch if that was what Thiokol recommended. To put it bluntly, it was simply not worth their while to launch if it was not safe. If the flight was postponed for a few more hours, it did not really matter, whereas if there was some accident or disaster, the financial and political cost would be enormous. So, Marshall was merely waiting for Thiokol to come back and restate their case.

However, the discussion at Thiokol was not going that way. The key point, as always, was redundancy. The engineers were not stating that the second o-ring *would* fail at low temperatures (if they had stated that, the discussion would have gone very differently). Instead, they were stating that *they did not know* what would happen at lower temperatures, which was a rather different position. They therefore could not prove (and did not attempt to prove) that the second o-ring would not seal. Instead, they argued that it might not or, to put it even more weakly, that there was no real data to show whether it would or would not. Now, of course, the social construction of shuttle safety and risk that had been painstakingly built up over the months and years counted against them. The engineers and managers had been so used to thinking of the o-rings as safe that it was hard to flip over and start thinking of them as unsafe. All the data that could be interpreted as showing the o-rings were unsafe could of course also be interpreted to show that they *were* safe, *as they always had been in the past.*

We have to spell out a point here we made implicitly in Chapters 3 and 4 (on Bayesian probability). Before any scientific argument can even begin, the debate has to be framed, in two specific ways: first by asking, "On whom does the burden of proof fall?"* and second by defining terms. Unless this (essential) social process can take place, then scientific debate simply cannot get off the ground. Before the teleconference, the cultural assumption had always been that the o-rings were safe, so the burden of proof had always been on those who wished to prove the contrary (that the o-rings were *un*safe); of course, these attempts had always failed. Now, literally on the eve of launch, some Thiokol engineers were suddenly attempting to turn this assumption on its head and were asking that NASA prove that the o-ring*s* *were safe* (at low temperatures). Of course, NASA could not do this, but Thiokol could not prove they *were not* safe, either. Like many, perhaps most, scientific experiments, the existing data were ambiguous and could be interpreted either way.

Second, the taxonomic category *safe* had to be defined, as well as the taxonomic category *low temperature*. Again, the devil is in the detail. A statement like "o-rings are not safe at low temperatures" could be agreed on by anybody, mainly because it is so vague as to be almost meaningless. What do you mean by safe? What do

---

* This is conditioned, in Bayesian terms, by how we set our "prior." And, this is a fundamentally cultural phenomenon, as should be made clear. We cannot see why the decision was made as it was without seeing that people instinctively think Bayesianly; that is, they assess future probabilities (in this case, of the o-ring failing) based on their cultural priors. What we had with *Challenger* was a classic Bayesian dilemma: two groups of people (NASA and Thiokol) with two different priors. In a Bayesian framework, the only way to sort out this problem is to run the experiment over and over until eventually (in theory) the probabilities of the two groups come together. Of course, on the night before the launch, this was not an option.

you mean by low temperature? The skeptics in the Marshall audience were perfectly happy to agree that the o-rings were not safe at low temperatures. The problem was that by the phrase *low temperature* they meant less than 30°F as in the Vitron management guidelines. To pick a figure like 53°F seemed to be ridiculous, especially since the shuttle had already flown at temperatures that were lower than this, and Thiokol had not raised any objections.

When Thiokol Senior Vice President Jerry Mason said the infamous words, "We are going to have to make a management decision," he did not meant that Thiokol were going to have to stop thinking about safety and start thinking about money and profits. He meant that the engineering arguments could go on all day, and he was right. The fact is that this is a classic situation of what we have called fuzzy logic: the o-rings were neither safe nor not safe; they were somewhere in the middle. The problem was, where? As we argued in Chapter 2, the answer is that some people will draw the line (i.e., impose a digital structure on an analog phenomenon) in a different place from other people.

The matter went to a vote between the four senior managers. Three of four recommended launch. After wavering, the fourth, Robert Lund (after being told to "take off your engineer's hat and put on your manager's hat"), also agreed to launch.

7.1.13 ISSUES RAISED BY THE *CHALLENGER* LAUNCH DECISION

What is the significance of the *Challenger* launch decision? We have argued that there are three main implications.

The first implication concerns methodology. Vaughan's discussion of the *Challenger* decision shows the benefits of thick description, which we have seen is closely allied to indexicality in ethnomethodology or contextualism in philosophy. Our temptation in the West is always to think that by being an outsider we can achieve objectivity, but these three concepts challenge that idea. We can certainly gain knowledge of the facts from the outside. We can understand little of what the facts actually mean without placing them in their *context*. That means to look at the various systems (not only local systems, but also organizational systems) of the people involved. It means to acknowledge the specificity of any situation instead of immediately looking to analogies to other situations, let alone attempting to create laws that will function in a nonspecific fashion. It means to immerse ourselves in the specifics in a situation until we perceive things, as much as possible, through the eyes of those who actually experienced them, that is from the inside not the outside.*

Second, we are so used to viewing phenomena in terms of laws or rules that it seems bizarre when they are not present. Therefore, whenever an accident happens, it seems clear to us that, if it was not wholly caused by merely technical failures, then there *must* have been at some point some form of rule violation. We tend, of course, to break these down into two classes: the inadvertent rule violation (the famous human error) and the deliberate rule violation, which we describe as a shortcut or even sabotage, corporate crime, or something of that sort.

---

* This might seem slightly vague, but we provide a more detailed description of how this fits in with our own methodologies in the last chapter.

In the case of *Challenger*, the Western mythology predisposes us to see the *Challenger* decision as an example of corporate crime: putting people after profits. Surely, we think, Thiokol must have broken some rule, law, or guideline in giving the go-ahead to the shuttle? Surely, *someone must be to blame!* So, we carry out expensive public inquiries, not only to see why the accident happened, but also to find out who was to blame and, if necessary, to punish them.

But, a Geertzian thick description of the events shows that there were no rule violations by Thiokol.* Instead, there were extremely complex social and organizational phenomena that interacted to produce an effect that nobody wanted or had planned. There is no evidence that if NASA had been more rule bound or that if everyone had adhered to the rules more strictly, the disaster would have been averted, although some tried to claim this after the fact.

This was a classic normal accident in Perrow's sense. It was the sense of redundancy (in other words, the engineering fix that was added to the o-ring to *increase* safety) that ultimately proved fatal. At the end of the day, the o-ring had always functioned adequately before, and the group could not shake the view that it would therefore function adequately again. The engineers had thought that the joint seal was a linear, closed system in which they understood how it would behave. They only started to consider that this was a nonlinear, open system that might behave in a completely unexpected way in certain conditions (or contexts) on the eve of the launch. By then, it was too late. Because of the extremely tight coupling of the shuttle system, by the time something had started to go wrong (in the launch itself) it was far too late to do anything about it.

The third point is the social effect on cognition. The decision to launch the *Challenger* was not something that took place in the head of an individual. It was a social process, and it was a social process that, to repeat, took place in a specific time, in a specific system. We do not find, in a lengthy description of these events, the neat little causal factors interacting with arrows pointing to boxes. Instead, we see infinitely complex sociotechnical systems, which are, strictly, unknowable by any individual, which produce complex and ambiguous scientific data, and which are interpreted by other social groups in different contexts, all with their own biases and interests [7].

7.1.13.1 Groupthink

So far this has been a fairly depressing chapter, and we should not let ourselves fall into the fallacy of retrospective determinism: the idea that some things *had to happen.*

---

* One could argue this is not strictly true, and that Thiokol violated the parameters of o-ring performance when it proceeded with flights when the first o-rings did not seal "enough." But, Thiokol of course thought this did not matter; these parameters were created to describe the old Titan 3 seals, not the newer shuttle design. From Thiokol's point of view, these rules were simply not relevant. In any case, experiments were carried out, and it was demonstrated that, despite the fact that the o-ring was in violation of (Titan's) rules/guidelines for performance, it still functioned adequately under conditions far more severe than those expected during a launch.

So, the question remains: what could have been done about the *Challenger* launch decision?

The key perhaps is to look at the third point again: the social nature of the event. Thiokol as a group created a social construct, a social construct of risk. Of course, some did not share this construct, but we have to see that these people immediately ran up against the catch 22 of groups: the further outside the group you are, the more likely you are to see the failings of the group, *but for that very reason*, you are less likely to be believed.

This should bring to mind the psychologist Irving Janis' classic work, *Groupthink*: "Groupthink refers to a deterioration of mental efficiency, reality testing, and moral judgment that results from in-group pressures" [8]. One of the key aspects of groupthink that Janis identified is the illusion of unanimity. To look at this further, we now turn to another example of system malfunction: *Columbia*.

7.2   COLUMBIA

Another day, another teleconference: this time it is January 21, 2003, at 7:00 A.M., and it is NASA mission STS 107, the space shuttle *Columbia* that is discussed [9]. As the shuttle took off, a piece of foam had broken off the shuttle's external fuel tank and had smashed into the shuttle's left wing. At this point, no one was clear what, if any, damage had been done to the shuttle. However, a group of engineers called the Debris Assessment Team (DAT) had used various computer simulations of the event to decide that there had probably not been any serious damage. At this meeting, the issue of the foam did not even get mentioned until two thirds of the meeting had completed; Linda Ham, who was in charge of the meeting, asked David McCormack (head of DAT) for an update. He told her that the engineers were not sure what damage had been done, and that it was not clear what (if anything) could be done to fix the damage (if any), but that his team was still working on the problem. Ham replied, "I really don't think that there is much we can do so it's not really a factor during the flight because there is not much we can do about it" [10].

The meeting continued. There was no further discussion. No one discussed the potentialities of the damage, whether it might be dangerous, extremely dangerous, or not dangerous at all. No one discussed telling the astronauts, asking them what they thought, or consulting with them (or anyone else) about what might be done about this problem (if indeed it was a problem).

There is no point here in going over the tragic sequence of events that led to the *Columbia* disaster. It is sufficient to say that masses of data came to light later that indicated that this was an extremely serious problem, and that if there was a suggestion that the foam might have damaged the protective tiles, then it should have been clear that the heat caused by reentry might lead to the destruction of the shuttle, which of course turned out to be the case.

What we are discussing here is this one teleconference, and the way it was managed such that an extremely serious problem that had to be tackled was transformed into a trivial problem, "which we probably can't do anything about anyway."

The tragedy, of course, was that this turned out to be false. In the subsequent investigation, it was discovered that NASA could have pursued two (at least) strategies that might have brought the crew back to Earth alive.*

What we are looking at here is a small group that made, disastrously, the wrong decision. Why?

There are, we argue, three main reasons. These are best seen not in the meeting of January 21, but in the meeting of January 24, when the DAT team presented their final report to NASA of their study of the foam strike scenario. By this time, after further research, the DAT team had become increasingly concerned about what might have happened to the shuttle's wing.

However, again (and we are tempted to state that in complex systems this is almost always the case) the data were ambiguous. NASA had decided that the foam strike on the shuttle was almost certainly not serious, and the only way the DAT could countermand that cultural belief was to obtain images of the impact point. However, because NASA had already decided that the foam strike was not serious, DAT had no apparent grounds on which to ask for the images they needed (another classic catch 22 situation). So, there were no hard visual data. Instead, the DAT team had to run computer simulations using an equation termed CRATER, which was, however, designed to mimic the impact of tiny meteorites and other space debris many hundreds of times smaller than the large piece of foam that in fact had hit the wing. So, they were not convinced that their simulations were accurate, and these doubts transmitted themselves to the Mission Management Team (MMT) in charge of this meeting.

However, the most important point was an important and interesting twist on the normal accident theme. In a classic normal accident, the accident occurs because the people in charge of the system are convinced that the safety net will function and stop the accident. In this case, however, the added twist was that the MMT did indeed think that there was no serious threat to the shuttle, and for reasons that are eerily similar to the reasons why *Challenger* was not thought to be at risk: there had been other, similar events that had *not* led to disaster. The operational definition of *risk* had drifted: the longer the shuttle had functioned without disaster, the more that people assumed this state of affairs would continue, and so on. However, as well as all these cultural beliefs, the MMT also assumed that even if there *had* been serious damage, *nothing could be done*. Here, they made the most amazing inference of all (which, to repeat, only seems amazing with the benefit of hindsight), that if there had been serious damage, the shuttle was doomed (which was true), *and that the crew were therefore doomed as well.*

This seems to have acted like a psychological block, preventing anyone from facing the reality of the situation. Again and again, as one reads the transcribed conferences one comes up against the "so what?" objection. Even if what the engineers were saying was true, so what? Nothing could be done. Better to say nothing and hope for the best.

So, in the meeting of January 24, what went wrong? There are, we argue, two things that basically prevented this group from working as it should have.

---

* The shuttle itself, of course, was doomed.

First, this was a *small* group. The meeting itself was not a small group. Actually, the meeting was so crowded that people had to stand outside in the corridor, a sign of how worried people had become as word of the results of the simulations crept out. The actual meeting that made the decision consisted of only the people on the MMT. In fact, this was not a meeting at all, merely an off-the-cuff discussion. The meeting itself (again at 7:00 A.M.) had consisted of the DAT outlining five scenarios based on different estimates of the size of the foam fragment and the impact point. Some of these scenarios were good, and some were bad. However, just as with the *Challenger* situation, the data were ambiguous (which provides evidence that the polysemous nature of scientific evidence is not the exception, it is the norm).

When the MMT met an hour later at 8 A.M., McCormack merely summarized the previous presentation, emphasizing that the results of the DAT scenarios were ambiguous and uncertain. No technical points were given (or asked for), and minority views (which stated the seriousness of the situation) were ignored. McCormack concluded by stating the following: "Thermal analysis does not indicate that there is potential for burn through. ... There is ... obviously a lot of uncertainty in all this in terms of the size of the debris and where it hit and the angle of incidence." Ham replied: "No burn through means no catastrophic damage?" McCormack answered: "We do not see any kind of safety of flight issue here" [11]. That, essentially, was that. Moreover, in the official minutes of the meeting (as everyone who has ever worked in an office knows, whoever writes the minutes essentially decides for posterity what the meeting had been about and what was decided) *made no mention of the debris strike.*

So, what happened here was a classic example of groupthink or false consensus. It was partly because the group was so small. Because it was small, it was easy for one or a few people to dominate it. Bigger groups tend to be more chaotic, and one can interpret this in a good way. People are less afraid (paradoxically) to speak out at big meetings: there is safety in numbers. People are less afraid to shout, raise their voices. In small groups, things are far more formal. It is not that the time constraints are less, but adherence is more likely.

Moreover, small groups are more likely (for simple reasons of numbers) to be cognitively homogeneous than large groups, and this is particularly the case in an organization like NASA. It is certainly true that in any group one needs consensus (and in fact should strive for it, as we have argued), but this needs to be a *genuine* consensus in which the issues are out in the open, and things are genuinely debated. Instead, due to time constraints and the smallness of the group, the actual discussion was over in about 30 seconds. This brings us to the second point.

Second, the group was hierarchical. It was clear that Ham was in charge. She set the agenda. She decided what would be discussed, by whom, and for how long. When she decided the issue was dealt with, it was dealt with. Again, this is a classic way to obtain groupthink. Instead of a free-ranging discussion, one got a quick, superficial discussion that was predicated on the (unspoken) idea that it did not really matter what they decided because they could not do anything anyway, and in any case things had been all right in the past and would probably also be okay this time. In other words, this was a small group that was governed top down by a small group of managers who found it easy to control the group and to maintain the hierarchy.

Information flowed from top to bottom easily enough, but it was much harder to make information flow the other way, especially as these managers had a knack of only hearing things that fit their cultural presuppositions (in this case, that the shuttle was fundamentally safe, and that even if it was not, nothing could be done).

7.3 CONCLUSION

What do we conclude from this chapter? We have stressed throughout this book the importance of the social context, and to analyze at such length how it can go wrong and lead to groupthink might seem to be a contradiction. But, when we mention the social in a positive way, we are actually pointing to something subtly different from groupthink, indeed, the opposite. What we are stressing is that groups might function better if they are (a) large, (b) cognitively diverse, and (c) nonhierarchical (as much as possible). By nonhierarchical, we mean a group in which information can flow freely, not only up, but also down. That is, information must flow up to managers, but equally, managers must keep staff informed and aware of what they are doing, and this information must flow down to be understood and reacted on, whereupon information in response to this must flow up again, in a feedback loop.

To some, this concept of the group having more insight than individuals might smack of mob rule. As discussed in Chapter 3, this is not true. Instead, a genuine, healthy social consensus can be built up from a crowd but, paradoxically, only if the crowd remains one of individuals. We have seen this in the discussion of reliability trials, for which the separate rooms and the like are meant to eliminate the hierarchy and groupthink that might otherwise destroy the group's efficiency.

Perhaps we can also see that, in the field of safety generally, we would be better to rely on groups (not self-appointed experts or groups of experts) of people who are socially and cognitively diverse, who are individuals, and who discuss matters in a nonhierarchical way. But, how could we find such groups? Or, how could we create them? We answer those questions in the latter half of the next chapter. Now, however, we must start to bring all the threads of this book together and to relate taxonomy theory to what we have just been discussing. To do this we must backtrack and return to our start: rules.

REFERENCES

1. Jost, J., Banaji, M., and Prentice, D. (Eds.), *Perspectivism in Social Psychology: The Yin and Yang of Scientific Progress*, American Psychological Association, Washington, D.C., 2004.
2. Geertz, C., *The Interpretation of Cultures,* Fontana, London, 1993, p. 6.
3. Harland, D., *The Space Shuttle,* Wiley, New York, 1998, p. 8.
4. Vaughan, D., *The Challenger Launch Decision,* University of Chicago Press, Chicago, 1997, p. 99.
5. *Funkhouser,* E., Three varieties of causal overdetermination, *Pacific Philosophical Quarterly,* 83(4), 335, 2002.
6. Adams, J., *Risk,* UCL Press, London, 1995.
7. Vaughan, D., *The Challenger Launch Decision,* University of Chicago Press, Chicago, 1997. Note that all subsequent references in the text to the *Challenger* crash are from this book.

8. Janis, I., *Victims of Groupthink: A Psychological Study of Foreign-Policy Decisions and Fiascoes,* Houghton Mifflin, Boston, 1972, p. 9.
9. The rest of this chapter is strongly indebted to the discussion of *Columbia* in Surowiecki, J., *The Wisdom of Crowds,* Abacus, London, 2005.
10. Columbia Accident Investigation Board, *Final Report*, Progressive Management, Washington, D.C., 1(6), 147, 2003.
11. Columbia Accident Investigation Board, *Final Report*, Progressive Management, Washington, D.C., 1(6), 161, 2003.

8 Rules and Regulations

8.1 RULES, PHYSICS, AND COGNITION

We are now approaching the end of the book, and we should begin to tie all the various threads of the argument together. First, however, we have to return to the theme of Chapter 1 and go over (one more time) the subject of rules and laws. Here, we attempt to show how Chapter 1 and Chapter 5 relate to each other, that is, how the idea of a law is the unifying concept that links technical rationality (TR) and cognitivism.

8.2 LAWS

Remember, the standard model of the laws of nature or laws of science goes like this: there are abstract laws of science/nature that have deterministic or causal power; they are not context specific (i.e., they are universal); they function at all times and at all places in the same way; we can discover these laws via abstract mathematical models or equations and then test that we have the right laws (i.e., laws that work) via experiment; this is how science progresses.

The problem with this view is that it can be shot full of holes. To begin, where precisely are these laws? Given their universal, timeless nature, it is tempting to give them Platonic, metaphysical status, but this clashes with a very large amount of empirical evidence that states that the universe was created 15 billion years ago in a big bang, and that the word *universe* should be interpreted in a very wide sense (i.e., meaning "everything"). So, time, space, and whatever metaphysical place where the laws of nature exist *would also begin in the big bang*. So, unless all our empirical data are wrong, the laws of nature could not, in fact, be timeless. There must have been a time before them.

But, if we posit the idea that the laws of nature, as we now understand them, are time bound, we therefore have to ask, how were they created? The answer must be (given this philosophy with which we have saddled ourselves) via other laws of nature (remember, this *must* be the case if *all* phenomena are caused by laws). This creation in itself must have been law bound, so, what caused this? More laws would be the answer. So, we are back in a situation of infinite regress, always a sign that something has gone seriously wrong with a concept.*

---

* Even if (as seems *highly* unlikely at present) the big bang theory is overthrown or replaced at some point in the future, we still have the problem of what are these laws? Logically, we must give them metaphysical status, or else it is difficult to see how they could be timeless, but this would seem to clash with the basic scientific assumption of materialism.

The concept of causality or determinism seems similarly incoherent. Again, we have a large amount of data from quantum mechanics that reality at the subatomic level is statistical, not deterministic. However, for the sake of argument, let us assume what seems *highly* unlikely at the moment, that some future scientific revolution might reinstate determinism and again look at the idea of the big bang. If we posit that everything is causal, again, we end up in a situation of infinite regress, that is, what caused the big bang? Before this first cause, we must have another cause that precedes it and so forth. Of course, infinite regress shows that we are not really explaining anything, but merely postponing the inevitable period when we must shrug our shoulders and say, I don't know.

But, there is a solution to both of these problems that we have tried to set out. This is to see the laws of science and causality as fundamentally *human* construction, linguistic (and numeric) ways in which human beings orientate themselves in the universe. In this view (which has precursors not only in the work of David Hume but also in the work of many other philosophers, notably Wittgenstein), we *create* the laws of physics and causality post hoc. We *use* the concepts of causality and causal laws of science to explain and understand, and we do this because this is our, human, way of seeing the world.

This view is termed *instrumentalism* in philosophy. Instrumentalists insist that the laws of physics, chemistry, and causality are pragmatic guides for action, ways of predicting events in certain contexts. Neither the laws of science nor causality itself, in this view, have metaphysical status. They are not true in some Platonic sense but of course can be true in a down-to-earth, pragmatic, useful fashion. I might create a law that every Saturday I do not go to work, and nine out of ten times it will be predictive and, hence, useful, but there *will* be exceptions (when one is working on a book like this, for example) [1].

This brings us to another key point: these framework laws are *context specific*. We can only tell whether they work when we define *in what context*. The laws relating to the speed of light will work in a vacuum (although actually they will never work exactly as predicted because these laws were only created in an as-if fashion to describe the behavior of light in an abstract mathematical world, not how light actually behaves in the real world) but will not work in the same way underwater or in air. We must define the context before we can decide which laws are relevant and how they will function (i.e., to predict things).

Of course, these ideas were controversial when they were first discussed, and in some circles they still are. However, in physics it tends not to matter whether one believes the laws of physics are metaphysically (Platonically) true; what matters is their predictive power. As long as they make reasonably accurate predictions, it does not matter that you think they are real (i.e., they correspond to reality) or merely instrumental; the fact is, they work in most situations.

8.3 PSYCHOLOGY

This view of instrumentalism did not become widely accepted until the 20th century (e.g., it is most associated with the Copenhagen interpretation of quantum mechanics, which is an instrumentalist theory). But, in the 19th century, when psychology was setting itself up as an empirical science, it still seemed to be simply common sense that science *was* the discovery of objective, context-free laws. Therefore, psychology *must be* the discovery of the objective laws of human behavior. So, this led to behaviorism, which was the self-proclaimed science of prediction and control (i.e., the discovery of deterministic laws of human behavior).

Behaviorism made other assumptions. It also assumed that a true science of psychology must concentrate on the individual. The reason for this was twofold. First, the behaviorists followed the view of Newtonian science in believing that science *had to be* reductionistic (i.e., reducing everything to its component parts; in this view, the individual human beings were like the atoms in atomic theory, i.e., the basic unit of analysis). Second, social psychologists (and sociologists) discovered that human behavior, even more than animal behavior or the actions of inanimate objects, was *situation specific*. (Remember data from, for example, the famous Stanford prison experiments showed that this was the case.) But, to deal with the social (i.e., the social *context* of behavior) challenged the reductionist (noncontextual) framework to which psychology was now committed. These data from social psychology could not therefore be assimilated to the new paradigm created, so psychology split into a psychology of the *individual* and a psychology of the *social*. However, in practice, this new science of social psychology remained a minority pursuit, a sort of intellectual ghetto. Real psychology was still presumed to be the psychological study of individuals.

This, it might be argued, was bad enough, but when the behaviorist paradigm began to break down in the 1950s and 1960s, the rejection of the social and the belief that science that did not find laws was not really science was still so entrenched that when the successors to the behaviorists (the cognitivists) began, as they thought, to look inside the mind/brain at cognition, it seemed logical for them to look for the *internal* laws that *must*, they thought, control human behavior (in the same way that the behaviorists looked for the *external* laws that controlled human behavior). Therefore, they started with the *presumption* that human behavior was law/rule bound and then went looking for evidence for this hypothesis and, of course, found it. You can always describe any behavior as law bound after the fact, and as Karl Popper pointed out, if you only look for evidence that supports your theory (and not evidence that might falsify it), then you can always prove your own theories, at least to your own satisfaction, whatever your theories might happen to be.

However, after this the cognitivists made an even bigger assumption: because it was possible to describe human behavior as law bound, this meant *that these laws actually existed in the human brain*. (Some went even further and claimed that these laws were more similar to the laws of formal mathematical logic than they were to the laws of physics [2].) Therefore, the ideology of the human as digital computer arose: human behavior must be caused by deterministic universal laws programmed into the brain in the same way that computer behavior was caused by deterministic algorithms programmed into its hardware. In short, cognitivists took over almost the entire metaphysical framework of behaviorism with only one difference: the laws of control of human behavior were now conceptualized as *real* laws (as opposed to merely conceptual constructs) that were programmed into the human brain in the same way that programs were programmed into the hardware of the digital computer.

This, then, is the link between Chapter 1 and Chapter 5: the notion of laws. We have strongly criticized this concept. We have shown that human behavior is best interpreted in a nonreductionist *systems* approach and is best seen as *highly* situation/context specific. We have argued in favor of an instrumental interpretation of not only the laws of science, but also those of causality.

Of course, this invites an important question: how do people actually think if not in a rule-bound way? It is very difficult for us in the West even to comprehend that other ways of conceptualizing cognition are possible, let alone true. We deal with that question in a moment. However, here we discuss how the concept of science as the discovery of objective laws had an impact on safety science. This happened in three ways:

1. When safety science began it was immediately assumed (following psychology) that safety science *must be* the discovery of objective (and therefore predictive) laws of, for example, accident causation. The corollary of this was that if such laws had not been discovered yet, then somehow safety science was not really a science, although it might become so at some point in the future. Even when this view was abandoned, or at least no longer insisted on, it was smuggled in "by the back door." So, for example, the context-specific nature of accidents and disasters tended to be downplayed; instead, there was a search for the one true model that would fit all accidents. This model would function as a de facto law for predicting accidents. The idea that it might be neither desirable nor in fact possible to deal with accidents in this way was not even considered.

2. First, the behaviorist and then the cognitivist or information-processing views of human cognition were embraced enthusiastically. The latter especially, with its emphasis on the human brain as digital computer, had two main corollaries: it emphasized (and overemphasized) the extent to which human behavior was determined by internal rules and it emphasized (and overemphasized) the extent to which human beings were atomized individuals. The influence of society, history, the group, and the social generally were downplayed. Instead of a branch of *sociology*, safety science became very much a branch of *psychology*, and cognitivist psychology at that.

3. Since safety science took over this view of science as the discovery of internal rules/laws in the brain, in a world which was rule/law bound, it followed inexorably that accidents must be caused by the *breach of a rule*. As we saw in the last chapter, it was generally felt that these rule breakages were of two distinct kinds: deliberate rule breaking (corporate malpractice or sabotage by individuals) or inadvertent rule breaking (human error). The idea that accidents might happen without any broken rules was rarely mentioned as a theoretical possibility.

It also led to a fourth corollary, which was rarely spelled out. We touch on this and then return to it in Section 8.5.2, but it is highly relevant to this discussion. It led to a highly specific conception of the nature of expertise and what was an expert. If one accepts that reality, fundamentally, is rule bound, then it follows from this viewpoint that to understand reality is to understand the rules or laws that govern reality. Moreover, the idea that these rules are abstract (i.e., not context specific; the rules were metaphysical or cognitive) means that these rules can be learned abstractly (i.e., in a classroom or from a book). Moreover, it means that anyone who knows these rules would be able to predict or control what would happen, and that this, therefore, would constitute expertise in any given situation. Finally, because these rules were not context specific but instead general or universal, it followed that to know these rules or laws in one situation was also to understand the rules or laws in any other context. The metaphor here is that of a pyramid: the higher up we go in the pyramid of science, the more powerful (i.e., abstract) the rules become (and the less context specific), with the most powerful and abstract laws of all belonging to theoretical physics, hence the search for the fabled theory of everything that would have total causal power *over everything*. The person who knew these laws would indeed know "the mind of God" (as Stephen Hawking put it), but more importantly for our discussion, the person would be a sort of "über-expert." To be fair, it was acknowledged that, *in practice*, this knowledge would be relatively useless because of its abstract nature. Instead, we would have to follow the pyramid down until we came to the merely pragmatic sciences (biology, zoology, and, below them, psychology, then sociology and anthropology), which would deal with nuts and bolts reality. Nevertheless, although the laws of these low-level sciences would not be as good as those of physics (given that the laws would become increasingly particular and specific the further down the pyramid one went), the basic idea that theoretical physics was to be used as a template for all the sciences was rarely questioned.

However, let us put this notion of expertise to one side for a few pages, and let us return to point 3: the notion of accidents always caused by some form of rule breaking. What is wrong with this view?

8.4 RULES AND REGULATIONS

We have seen one problem with this view in the discussion of the *Challenger* decision (Chapter 7), for which no rules were broken. But, we can look at this notion more deeply. The philosopher Ludwig Wittgenstein pointed out in his great work, *Philosophical Investigations*, that "This was our paradox: no course of action could be determined by a rule, because *every course of action* can be made out to accord with the rule" [3; italics added].

What did he mean by this? What he was getting at is something we touched on when in the discussion of hermeneutics: rules (or symbols, discourses, or behaviors) do not simply pass into our brains, like a program gets put into a computer. Instead, they are interpreted by human beings, and they can be interpreted by different people in different ways at different times. Therefore, in an organization, even when management puts down what it thinks are crystal clear rules for behavior, these can always be interpreted in different ways. This is *not* just a matter of stupidity of the workforce or willful misinterpretation of the rules (although, of course, it might be). We return to this point shortly.

8.4.1 CASE-BASED REASONING

We have so far given the impression that perhaps we are merely (sophisticated) behaviorists: perhaps we do not believe in cognition? That is not true. But, in that case, what is our model of how people actually think? We have provided one answer to that in Chapter 5. However, let us look briefly at another model of cognition that might be useful in this particular discussion.

Roger Schank's concept of case-based reasoning (and the related idea of *dynamic memory*) was one of the first major challenges to cognitivist orthodoxy in the 1970s and 1980s [4]. Case-based reasoning is very simple to describe. It is the theory that, instead of being rule bound, cognition is *case bound*. So, for example, according to cognitivism (rule-bound cognition), I can walk through a door because I have a stored algorithm that tells me how to do this: the algorithm tells me to put leg 1 in front of leg 2, then repeat, then lift arm, then open door by turning knob, and so on; that is, *the brain is a digital computer.*

For Schank, on the other hand, one cognizes via building up a library of cases. In other words, when I go shopping, I manage this because I have been shopping before, and I remember these cases. If there is no specific case to draw on (if I have never been to a specific shop before), I adopt the nearest possible case (going to similar shops) and reinterpret and reimagine this case with relevance to my own situations. In other words, the more shops I have gone to, the more of an expert I will be at going to the shops.

Coupled with this is the idea of dynamic memory, the idea that instead of being inert, memories are (at least in some cases) stored in the form of meaningful *stories.* In other words, cognition is largely the art of creating stories from cases. Instead of remembering algorithms, I will remember the time I went shopping and I bumped into my friend John Smith, and we discussed the game on Saturday or whatever.

The key difference between this and the cognitivist or algorithmic view of cognition is that "In order to understand the actions that are going on in a given situation, a person *must* have been in that situation before" [4; italics added]. I learn how to go to a restaurant, for example, by going to a restaurant. The more restaurants I have been to, the more I will know how to do it.

Now, of course, there is a first time for everything, and there will always be the first time I go to a restaurant. But, in this situation I *adapt my knowledge of previous cases.* For example, I have not been to a restaurant before, but perhaps I have been to a McDonalds, a fish-and-chip shop, or some other kind of shop or store analogous to this experience. The closer this situation was, the more adaptable my case knowledge will be.

Therefore, the more things I have done, the better I will be able to do other things because I will be building a dynamic library of cases. This has two corollaries: *life is basically about learning* and *learning is highly context specific.* So, are Schank's theories useful?*

---

* Sharp-eyed readers will have noticed that although Schank's ideas are *not* cognitivistic, they are an information-processing theory, and they are open to homuncular objections. However, the key point of Schank's theories is that they prioritize experience and learning; this is the aspect in which we are interested. We are not so interested in whether they are true in some Platonic sense.

There is a considerable amount of evidence that they are. For example, Klein and Calderwood [5], in a study of firefighters, found numerous examples of using previous experience in the form of specific instances to guide current or future actions. For example, in one case the fire commander remembered that hot tar running off a roof had ignited a secondary fire in an earlier incident and used this learned information to prevent a recurrence by putting water on the hot tar. In another case, the commander recalled administrative problems in an earlier forest fire in which there were two teams working at the same time, so he set up a second camp for the second team, therefore preventing this problem.

However, this simplifies the situation because it makes it sound as if these men always referred back to one specific case. In actuality, after decades of experience, in real-world situations they made reference back to *many* previous cases; their experience consisted of these cases. In any event, as Klein and Calderwood stated: "We have concluded that processes involved in retrieving and comparing prior cases are far more important in naturalistic decision making than are the application of abstract *principles, rules* or conscious deliberation between alternatives" [5; italics added]. Therefore, the expertise of these men consists of simply having done something a lot.

This leads to a radically different conception of an expert: an expert is someone who has done something a lot. If I have driven a lot, then I become an expert driver. If I walk to my local shops, then I become an expert at walking to my local shops, and so forth. This is backed up by experimental data, which tend to predict that the best predictor of performance is simply practice, and this is as true of cognitive activities as physical activities [6].

8.4.2 RULES AND REGULATIONS (REDUX)

We have seen the links between cognitivism and the standard view of accident causation. However, to repeat, apart from the fact that both share the reductionist philosophy of science (discussed further below), it is their emphasis on rules that unites them. Cognitivists assumed that there were internal rules inside the human head.* In the same way, the classic view of accident causation assumes that organizations must be (i.e., should be) rule bound as much as possible and that accidents involving human beings must involve a rule violation at some point (the infamous human error).

These two views are clearly related. If human beings are rule-bound cognizers, then it follows that organizations (made up of human beings, after all) must also be rule bound as much as possible; after all, that would only be natural. Therefore, malfunctions in the organization (accidents, emergencies) must be the result of someone breaking a rule, either intentionally (a violation) or unintentionally (human error).

---

* To repeat, they believed this because they believed that if human cognition was not rule bound, then no science (i.e., science like physics) of psychology could exist. So, they began with the assumption that these rules *must* exist and then went out looking for evidence for them. Compare and contrast the following: the Loch Ness Monster *must* exist; it is just a matter of finding the evidence.

But what if this is not true? If human beings are case-based reasoners instead of rule-bound information processors, then this analogy fails. Moreover, we have the empirical evidence from Chapter 7 of accidents occurring *without any (major) rule violations.*

This calls into question the pyramid concept of management organization discussed in Chapter 1 (and which is obviously linked to the pyramid view of science; see below). In this, managers are at the top of the pyramid, and staff/workers are at the bottom. The key task of managers is to make the rules. It is the role of staff to obey the rules. If they do not, then the organization suffers, and there is increased risk of an accident. The role of a safety manager is to force the staff to obey the rules. What if this view is wrong, not just slightly wrong, but fundamentally and completely misconceived?

To begin: who says that accidents are always caused by rule violations? As discussed in Chapter 7, this does not seem to be true. Furthermore, safety rules can be *wrong.* Ultimately, they do not descend like manna from heaven but are written by real flesh-and-blood, situated people, whose take on a situation is, like all other forms of behavior, situational not universal.

A seemingly trivial example given in Woolfson et al.'s *Paying for the Piper* demonstrates this. "A pipe-fitter on the Amoco Montrose [an oil rig platform] was given a 'hot-work permit' and instructed to cut into a length of pipe. The pipe contained potentially lethal explosive gases, but the worker did not proceed with the job … and thereby forestalled what could have been a major accident." Woolfson et al. argued that the accident was prevented "only because the worker was prepared to exercise his *initiative* and adopt what he felt to be a safe work practice. The individual worker contested an already managerially approved task assignation on his own '*tacit knowledge*'" (italics added). The authors continued: "Hierarchical organisations tend to concentrate power and therefore 'expertise' as part of managerial control … [but] for 'risk valuation from below' to be effective, the workforce needs to be informed. Much more than this they need to be involved in the process of risk assessment as fully legitimate active participants" [7].

We can look even more fundamentally at this problem. After all, perhaps those were merely "bad rules," the bad apples that spoiled the rule barrel. Return to Wittgenstein's statement: what he was saying is that *all* rules are inherently ambiguous, and that there is no real meaningful distinction between rule breaking and not rule breaking. Instead (and this is central to a hermeneutic approach), what we are really talking about are specific actions that can be *interpreted* as rule violations by specific people with particular motivations in specific situations.

8.4.3 SEMCO

Let us now look at what the great Brazilian entrepreneur Ricardo Semler said about his firm, Semco. Semco was a typical Brazilian manufacturing firm until Semler radically democratized it, creating a system in which, for example, managerial staff decided on their own pay and bonuses, workers set their own productivity targets and schedules, and workers voted for their own bosses. The result is that Semco is now one of the fastest-growing companies in Latin American.

Here is what Semler said about rules and regulations. "We tried to write new rules ... but at every turn we found ourselves wading into a swamp of minutiae." After a bad experience trying to reorganize the car fleet, he said:

> We quickly concluded that some departments were better not created *and some rules were better not written.* Common sense would be the best alternative. ... *We were trading rules for common sense.* And that is the system we have today, which is barely a system at all. When you get a company car at Semco, you can do anything you want with it. If you have a friend who is a mechanic, have him take care of it. [italics added]

Semler continued:

> Where do all these rules come from anyway? They were ... an unhappy by-product of corporate expansion. How does an industrial giant act as it grows? First management concludes that a company cannot depend on individuals. ... Next thing you know, committees and task forces and working groups are spewing out procedures and regulations and stomping out individuality and spontaneity. In their quest for *law, order, stability and predictability,* corporations make rules for every conceivable contingency. Policy manuals are created with the idea that, if a company puts everything in writing, it will be more rational and objective. ... and so it became accepted that large organizations could not function without hundreds or thousands or tens of thousands of rules. [italics added]

Semler then gave an example of why this approach does not work:

> Semco had a particularly complicated set of rules on travel expenses. Our auditors often spent hours arguing over whether someone on a business trip should be reimbursed for movie tickets. What then about theatre tickets? What do we do if an employee went to a show that cost $45? Or $100? And what about calling home? How often should the company pay? Was a five minute call reasonable? What if the employee had, say four children? Are 75 seconds per child sufficient? [8]

What Semler is pointing out here is what Wittgenstein also pointed out: rules are *inherently* ambiguous, and therefore they often *mean what you want them to mean.* They appear to provide objectivity and controllability, but in fact they produce waste, inefficiency, and a bureaucratic culture. They also move power up the corporate structure to managers. That is why managers like them, but it guarantees that the company will tend to become sclerotic and ignores the bottom-up innovative practices that bubble up from the bottom, from the *real* experts.

8.4.4 RAILTRACK

To give another example of this, let us look at the Railtrack rule book from the old U.K. railway system. This is important because after (depressingly frequent) major rail crashes in the British railway, there is usually a public inquiry that discovers that there have been frequent rule violations.

Here are some rules taken at random from Railtrack rule book Section Hiii, 30, rule X. 6. 7. 5. (it should be noted that this is from the master rule book; there are also 13 separate rule books as well for specific varieties of staff): "If there is a failure of the door operating controls the following procedure must be carried out: [In the event of] failure at one position, if possible the Guard must operate the doors from another position on the same side of the train."

Now let us stop here. Note the use of "if possible." Who decides if it is possible? The driver? The signaler? The guard? What if they have a difference of opinion, and one thinks it is possible, and another does not? Is this a rule violation? From whose point of view?

Continuing from the rule book:

> Circumstance, failure at all positions. The guard must ... use the passenger's emergency door release facility by breaking the glass cover to release one door: manually open and close the door, advise passengers of the arrangements, advise the driver. The driver must then report the circumstances to the signaller immediately, [and] act in accordance with the instructions given (which will normally include proceeding forward to the locations where one of the arrangements in clause 2.2.2.4 of Rule Book Section H (part 1) will be carried out).

This is not a parody. This is a real rule from a real rule book. If an accident resulted, then no doubt the events leading to it would be described as rule violations, but does that phrase actually *mean* anything in this context? Look at that rule again, and now consider some hypothetical questions. What happens if some, but not all, of the door operating controls fail? There is no rule to cover this. Then, we have the order in which the rules are presented. The guard must use the emergency door release facility first. If that is not working? The rules do not cover this eventuality. In any case, he must do this first, then contact the driver. Say, he contacts the driver first. This would be a rule violation, but it might be common sense in a safety-critical situation. Then, the driver must report the circumstances to the signaler immediately. We must remember, of course, that immediately here cannot be defined literally. Instead, it means as soon as possible (obviously), and then again, we are stuck back in the same situation as above: who decides when it is possible, that is, what *immediately* means? The signaler? The driver? The guard? Of course, the signaler must reply and has his own rule book. But, say (as seems highly likely) he has his own interpretation of the rules? Who has priority? And, going back, how does one "advise passengers of the arrangement"? What constitutes advising them? Say the guard decides in a safety-critical situation that this would cause panic? Is this a rule violation?

Therefore, we can rewrite Wittgenstein's point as follows: this was our paradox: no course of action could be determined by a rule because *every course of action* can be made out *not to be* in accord with the rule. It can always be argued that *any* activity is a rule violation from some point of view. The concept explains everything and therefore nothing.

8.4.5 THE SOCIOLOGY OF RULES

Finally, to give an example of this in practice, the following is the sociologist Anselm Strauss's description of the use of rules in a hospital examined as part of an empirical study.

> In (the hospital) ... hardly anyone knows all the extant rules, much less exactly which situations they apply to for whom and with what sanctions. If this would not otherwise be so in our hospital it would be true anyway because of the considerable turnover of nursing staff. Also noticeable to us as observers was that some rules once promulgated would fall into disuse, or would periodically receive administrative reiteration after the staff had either ignored these rules or forgotten them. ... Hence we would observe that periodically the same informal ward rules would be agreed upon, enforced for a short time, and then be forgotten until another ward crisis would elicit their innovation all over again. As in other establishments, personnel called upon certain rules to obtain what they themselves wished. ... Elsewhere too all categories of personnel are adept at breaking rules when it suits convenience. ... Stretching the rules is only a further variant of this tactic ... *hence the area of action covered directly by clearly enunciated rules is really very small.* [9; italics added]

Clearly the idea that organizations are, or ever could be, fundamentally rule-governed entities must be questioned, as should the consequent view that to make an organization safe (i.e., give it a good safety culture), it is merely necessary to set out the rules clearly and ensure that staff follow them.

It should be noted here that we are *not* arguing that all rules should be scrapped, and that we should all merely rely on what the philosopher Michael Polanyi called *tacit knowledge*, important though that is (as we saw in the Amoco Montrose situation) [10]. What we are arguing is that the presentation and the *meaning* of the rules in the rule book should change. Instead of an interminable (and probably contradictory) list of dos and don'ts, we are suggesting that rules should become broader and more general, common sense descriptions of the way that task performance has evolved. Therefore, instead of rules with the hidden implication "do this, or you will be fired," they should perhaps be offered more as "these are some of the methods we have developed of doing these particular tasks here, and we have found them useful."* It might be useful to follow these rules. Alternatively, one may have to break them to do the job in hand more efficiently and safely. The decision, ultimately, comes down to that complex mixture of culture, learning, and attentiveness to the situation we term (as Semler pointed out) *common sense*.

8.4.6 TRAINING

Schank's concept of case-based reasoning also has implications for training. Instead of simply passively receiving (ambiguous and possibly incoherent) rules, perhaps staff could be encouraged to *study cases* instead? What we are suggesting here is that there might be social (i.e., group) workshops in which staff could talk over specific cases from the past and discuss these: what would we have done? What can we learn from this? What did the staff involved do right? What did they do wrong? The results would not be more rules but instead a deeper understanding of the organization and their own role in it. Moreover, it would result in them putting forward their own views of how they actually proceed based on their own experience.

---

* In other words, there is the need to see rules as heuristics, fuzzy recommendations, rather than deterministic algorithms.

Remember, in the view of reflective practice, what must be avoided is that the technical expert tells them what to think. Instead, we act under the assumption that these people have their own experiences, their own thoughts and views that may well be as interesting and useful (or more so) than our own, and that we, as well as they, might learn something from the resulting discussion if we just learn to listen.

8.4.7 EMERGENCE

One of the key themes that have run through the book concerns *emergence*. It should now be clear that this unites systems theory, neural nets/connectionism, and this view of expertise. In all three fields, things work *bottom up*, not *top down*. This, of course, is also the classic message of complexity theory, which should really be seen as a part of systems theory. Expertise should be seen as an emergent property of the simple actions of doing the thing you do, over and over again [11].

Again, the *organization* should also be also seen as operating bottom up rather than top down. People on the bottom perform the tasks at the cliff face: they are, in a very real sense, the experts, not the managers at the top, who have merely rule-bound abstract knowledge. Remember, according to Schank, to know how to do something you actually have to do it [4]. There is no other way. And, to become good at something, you have to do it a lot. The complexities of organizational behavior and of human beings are an emergent phenomenon of us doing things over and over again, slightly different each time, which gives us these cases we need for learning.

8.4.7.1 Science and Emergence

With this knowledge, we can now tie this in with what we learned about the sociology of science from Chapter 1. The classical view of science is the top-down view par excellence. In this view, science is a pyramid with theoretical physics at the top, then applied physics, then chemistry, then biology, and so on, down to the human sciences. Theoretical physics is at the top, of course, because its laws are the most abstract and, well … theoretical.

We are arguing that this pyramid structure (and the analogy of the pyramid structure of an organization is no coincidence) no longer makes any sense. Instead, as Nancy Cartwright suggested, we should see the sciences as more like a patchwork quilt, with various context-specific strategies and heuristics that have evolved to be used in specific circumstances [12]. As Karl Popper once pointed out, for any description of a subject of any complexity, more than one scientific law (perhaps three, four, or more) will be needed. These, of course, might come from physics, chemistry, or biology; the precedence we give each one will depend on the nature of the investigation [13]. In other words, instead of moving up to a smaller and smaller number of macrotheories,* we are suggesting that psychology, sociology, and safety science are, and should be, a collection of *microtheories*, small-scale, domain-specific, instrumentalist theories.

---

* By a macrotheory, we mean a classic mathematized law of the type we are used to seeing in theoretical physics.

There is one final twist. Remember from the discussion of the space shuttle incidents that any given experiment is *overdetermined*. That is, for any given experiment, more than one hypothesis might explain it. This is an issue in physics and engineering, as we have seen, but it is even more of an issue in anything involving human beings because of the greater complexity of the subject. Therefore, we should view the landscape of the sciences not only as a patchwork quilt, but also as one we must move over again and again, continually returning to the same territories but each time from a slightly different angle in a slightly different way. Because of this, there will be no end to this mapping. We can always find a new viewpoint, a new way of looking at the territory that will lead to a new research program.

For example, it might seem that when discussing Schank's theories we are saying that Schank's theories are *the* explanation regarding how people learn. In fact, there are many theories of how people learn, such as Bandura's social learning theory, which posits the idea that human beings learn by watching other people's behavior and copying it [14]. Which is correct, Schank's theory or Bandura's?

What is suggested here is that, as long as these theories do not actually *contradict* each other, they can *complement* each other. *We do not have to choose.* In some circumstances, when we want to make specific predictions or model behavior in a specific way, Schank's theories will be pragmatically more useful. In others (e.g., when we are concentrating on the social aspects of learning) Bandura's will be more useful. In other words, we see Schank's and Bandura's theories as microtheories, not macrotheories. Psychology, safety science, and the social sciences generally, we are arguing, should be seen more like a patchwork quilt than a pyramid, with an ever-expanding network or web of microtheories, endlessly criss-crossing the same territory, and slowly expanding outward as more is discovered.

In short, will safety science ever become a true science? If one means by *science* "slotting into the pyramid" somewhere above psychology but below biology the answer is clear: safety science *will never become a science in that sense.* Instead, it must take its place on the map; another territory to be explored that will be viewed in a variety of different ways as culture (and the surrounding scientific environment) alters.

Now, we are aware that this is going to be difficult for some people to take. For decades, workers in the field of safety science have been waiting expectantly for some Newton or Einstein figure to provide a paradigm that will unify the field and give safety science the rigor that (we are told) exists in theoretical physics. What we are arguing here is that this will never happen because the presuppositions on which this view of science rests are false.

But there is no need for despair. On the contrary, this new openness to methodological pluralism should be a source for optimism. Again, we must emphasize the pragmatic, fragmented nature of the subject. What we are looking for is what works: what will prevent accidents, what will improve safety. We have to fit the tools to the job. We have in this book stressed the usefulness in our experience of some tools (e.g., taxonomies) that we hope will be useful for safety scientists. We are not suggesting that safety scientists should stop performing experiments. We *are* suggesting that the technically rational approach that has proved useful in safety science (and psychology) in the past has now run its course, and we should look for new methodologies and philosophies. We are not suggesting that some new paradigm of the sort we have become used to seeing in the field should replace TR in the same way that behaviorism was replaced by cognitivism. Instead, we are arguing for a new era of *methodological pluralism*, with the metaphor of the patchwork quilt replacing the pyramid. We are providing some new tools for the toolbox, but we are not at all saying that from now on everyone should do things our way or that there should be *any*, one, single approach or methodology in the field. Instead, there are specific problems with specific tailored solutions, and the method should fit the problem. There should be as many approaches as there are problems to solve.

8.5 TECHNICAL RATIONALITY

Let us move back swiftly from the abstract to the concrete and look back at what we should actually be doing about safety. This brings us back to Donald Schön's view of reflective practice because it was precisely Schank's view of expertise (i.e., the case, *experience* based) that Schön's philosophy presupposes.

We can begin to look at the two related questions: given all these factors, what should a safety manager actually do? What has this got to do with taxonomic theory? First, let us look over what Schön said. Then, we look at a specific example of reflective practice in action.

8.5.1 REFLECTIVE PRACTICE

To put it bluntly, reflective practice was developed to deal with a world of cases, not laws. In other words, in many (perhaps most) situations we are dealing with a universe of one: a specific instance that can be compared to other situations only by analogy. Although Schön's work has tended to be used in education, he made it clear that it has a far more general application [15]. For example, he discussed the development of the transistor in the 1930s and 1940s and showed that if one looks at what the engineers and scientists involved *actually did* (as opposed to what they were supposed to have done), it accords much more closely to the case-based reasoning we have examined as opposed to the use of predictive laws or rules.

If this is the case, then, by a process of analogy we can question the role of the expert as that word is used in what Schön called technical rationality. As Schön put it:

> Within the dominant tradition which has grown up over the past 400 years, the professional's claim to extraordinary knowledge is rooted in techniques and theories derived from scientific research undertaken for the most part in institutions of higher learning. The status of professional experts, their claims to social mandate, autonomy and license are based on the powerful ideas of Technical Rationality and the technological program. [15, p. 288]

It is this paradigm that is almost invariably used throughout safety science. As we saw, it was first stated explicitly by Heinrich [16], and even though his work has been widely criticized, those who criticize it normally remain within the same technically rational framework. In other words, accident investigators, safety scientists, safety managers, and others are experts: they have access to (or are attempting to access) the rules or laws of accident causation to predict and control accidents. Again, those who have argued against this viewpoint (e.g., James Reason [17] or Jens Rasmussen [18]) are "beating the door down with one hand and holding it up with the other." They insist on the value of the social, on the fact that organizations are (metaphorically) like biological organisms, should be viewed in that light, and so forth. But, because they have tended to use the psychological framework of cognitivism (the technically rational approach to psychology par excellence), they tend to restate the values of TR even as they argue against them.

However, if one dismantles the framework of TR from its roots, this will inevitably have an impact on the way safety scientists and engineers see themselves and therefore how they do their jobs.

8.5.2 THE NEW VIEW OF AN "EXPERT"

Schön argued that experts are not those with knowledge of abstract laws but, on the contrary, are those with experience of *specific cases* [15]. When one has done a thing 100 times in a 100 different contexts, one will know how to do it better than after one has done something once, and it is irrelevant whether we are talking about scuba diving, cooking, working on a physics project, or whatever. Therefore, instead of being an expert who has *possession of knowledge* (in the form of predictive rules or laws) that he or she will impart to a client (or workforce, or management, or whoever), the expert has a casebook of experiences to *share* with the client. But, ipso facto, by the same logic this must acknowledge that the client also has had experiences, some of which might be relevant. The client can learn from the expert, but the expert might also learn from the client.

Moreover, the viewpoint of reflective practice also illustrates the project-specific nature of any task for which the expert will provide help. A group of people are brought together to do *this* project at *this* time. By analogy, other cases from the past might be useful, but equally, they might be misleading. *Not* to look to the past at all would be madness, but we should not be misled: each incident, accident, or situation is fundamentally different from all others, and we should draw inferences from past cases with care.

In our own case, for example, when we are brought in as experts to build a coding system of safety-related data, we are always aware that this is a specific project unlike the others. Therefore, we are extremely wary of off-the-shelf software packages or techniques that can be "adapted" for specific situations. Likewise, taking models of an accident (as though it was possible or meaningful to create a single model of *all* accidents or incidents) might again lead to ignoring or downplaying the specific aspects of *this particular* situation.

Hence, when we are asked to build a taxonomy that will organize a database or set up a reporting system, even though we are using our expertise as creators of taxonomies (because we have done this a lot), we always pay attention to the project-specific nature of each database and situation. We argue strongly that no off-the-shelf taxonomy will ever be able to fit all situations. There are simply too many variables. Some companies want to know more about human factors, others more about technical factors. Some have the money for intensive training in the use of a taxonomy/coding system; others do not. Some systems are confidential; others are not. Different data sets have different demand characteristics that mean taxonomic hierarchies should be more or less fine grained. Moreover, the use of the system has to be tailored to the needs and desires of the people who actually have to use it. Again, all these factors necessitate situation-specific taxonomic design. There will never be just one model of an accident, and there will never be just one taxonomy that can be adapted for all situations.

Finally, the reflective practitioner tradition, as the name suggests, privileges practice over theory, experience over knowledge of abstract laws learned in a university. Again, in this view expertise is simply the result of having had experience of a large number of cases in the real world.

This sounds good, but what does it mean in practice? Schön gave an example of the reflective practice approach working in a safety context by an engineer named Dean Wilson:

> He (i.e., Wilson) had been asked to help solve a problem that had arisen at the teaching hospital in Cali. Doctors there had become aware of a high rate of error in the administration of drugs to patients. They blamed the nurses and believed that the solution was to put all of them through a master's program, but they knew that this would be time consuming and costly. Wilson began by mapping the 'flow process' involved in the conversion of a doctor's order to an administered drug. He found an error rate of 33%, compared to an average error rate of 5% in the United States: and by applying the method of binary search (that is, by measuring error rate at the midpoint of the process and then at the midpoints of the remaining segments) he found that errors were evenly distributed throughout the whole process. He presented results to a meeting of doctors, nurses and orderlies, and asked them for their ideas about ways to reduce errors. In the corridor, in full view, he placed a chart which showed the previous week's error rate at each stage of the process. *As the staff found ways to reduce errors*, the changes were recorded on the chart. Gradually, over a period of three months, error rates dropped to the U.S. norm. [19; italics added]

As Schön wrote: "The doctors and their staff … functioned both as problem solvers *and* as implementers of their own solutions" [19; italics added].

The difference between this view of expertise and the view of TR can be stated clearly. In a TR framework, the expert would come into the situation and *tell* the staff what the problem was. The expert would already know what caused accidents, would present the staff with models of accident causation and human error, and *tell* them that this was the expert's view. Then, predicated on these views, the expert would study the situation but within the ideological framework created by TR. Eventually, the expert would discover what caused the high accident rates, and then, on his or her own or possibly in consultation with management, a plan would be devised that would be unveiled to the workforce, who would then be *told* what to do.

In the view of a reflective practitioner, on the other hand, it would be presupposed that the doctors and nurses were the experts. After all, they did the job. They had access to the library of cases that constituted their expertise. Not only did they have this experience, but also (given that they did not want patients to die) they were fully motivated to improve safety. However, what they lacked was a method to reflect on their own practices to improve them. This Dean Wilson provided. Therefore, he was a facilitator of the staff's *own* attempts to improve their *own* work practices. He did not at any point pretend to have more knowledge than the staff themselves. What he did know how to do was to help people help themselves, and with the notice board, he achieved that.

From the point of view of systems theory, we can look at this as a classic example of a system *self-organizing*.

8.6 SELF-ORGANIZATION

It is hoped we are now at the point at which the whole thrust of this book should be clear. One of the key things we emphasize is the creation of taxonomies in many different circumstances. One of the key purposes for which we have used them is to create reporting systems for accidents and incidents. We have done this with CIRAS (Confidential Incident Reporting and Analysis System for the U.K. rail system), SECAS (Strathclyde Event Coding and Analysis System, a nonconfidential system for the nuclear industry), and CARA (Confidential Accident Reporting and Analysis, the nationwide confidential reporting system for the railway system of the Republic of Ireland), and in numerous other situations. Now, the standard reason for such systems is to see through the language of the reports to see the real causes and the cognitive architecture that lies behind.

We have rejected that view. So, what is the point of such systems? Based on the Schön's example, we would argue that such systems enable organizations to self-organize. In other words, data are sent up to the management, who then feed information down to the workforce, creating a virtuous circle of reflection back on actions that can, as in the case above, lead to reductions in accidents. In the case above, we must remember that what Wilson essentially did was to set up a reporting system (like CIRAS or SECAS) that enabled personnel to reflect on their practice; this led to a lower accident rate. Therefore, this is an example of a system self-regulating. More than this, it is an example of the group knowing more than the individual. The cognitively diverse group (of doctors, nurses, etc.) knew how to self-regulate its own system, knew in fact more than the solitary expert.

This (finally) brings us back to taxonomy theory.

8.6.1 EXPERTS

What is the *real* problem with the view of the expert we have looked at here, the view of TR? The real problem is the most fundamental: (technically rational) experts *get it wrong*. For example, remember the Rasmussen report from Chapter 3, generally considered "the most complete hazard analysis ever accomplished" [20]. This report pioneered the idea of probabilistic risk assessment in terms of nuclear power plants (NPPs). It analyzed what was thought to be all possible ways in which a NPP could malfunction and then quantified them in terms of the likelihood they could happen. It concluded that the probability of a core melt was about 1 in 20,000 per reactor per year, and that the probability of 1000 or more fatalities was 1 in 100 million [21].

Obviously, the Rasmussen report stood or fell on the extent to which it could predict every single thing that could go wrong with an accident. However, one of the key elements of plant safety in an NPP is the emergency core cooling system (ECCS). In its assessment of the ECCS, the Rasmussen report failed to hypothesize that when the ECCS was engaged, something *else* might go wrong with the plant. However, in 1981 and again in 1983 (at Browns Ferry and at Salem, South Jersey, respectively), safety events occurred; in both, it became highly plausible that the ECCS could have been overwhelmed in the midst of other events (i.e., because something else went wrong when the ECCS was engaged).

Our wish is not to denigrate experts here, although we could quote many, many other circumstances in which the experts also failed to get it right. It is also much broader than a simple critique of probabilistic risk assessment, although we prefer Bayesian approaches. Our point is more fundamental. If you remember the discussion of laws in Chapter 1 (and earlier in this chapter), you will remember the debate between the realist and the instrumentalist view of the laws of nature. The key point is that, in terms of practical physics, it does not really matter whether one believes in the laws of physics as abstract metaphysical laws; the fact is that they work. Spacecraft (usually) go up in space, planes fly, cars get from Point A to Point B, computers operate, and so on. All these depend on the functioning of the laws of physics, chemistry, electronics, and so on. The laws of science are not like Santa Claus: they give you the goodies whether you believe in them or not. They function and are seen to function because they are predictive.

As we have seen in the examples in this chapter, it is not clear that safety scientists are anywhere close to creating such predictive laws, and (apart from pious wishes) there is no indication that they will create such laws (or models) at any point in the future. But the prediction of events is, as it were, as a guarantor of the expert nature of, for example, experts in physics or chemists or engineers who work in fields derived from physics or chemists. We accept as experts people who design airplanes and cars because their expertise is open and public and therefore checkable. Either the engine works, or it does not; either the plane flies, or it does not.

There is no similar guarantor for a safety expert. Safety experts, in practice, cannot in fact predict accidents. They cannot (as the Rasmussen report indicated) tell us what will happen in an NPP. They cannot even tell us what cannot happen (see also Chapter 6). So, the top down view of expertise in the safety field becomes questionable. The lone (usually male) expert, skilled in rationality and whatever else it is that safety experts are skilled in, seems to become an anachronism when he or she cannot provide the (accurate) predictions that back up the expertise of, for example, the physicist or chemist.

8.7 THE SOCIAL VIEW

As an alternative, we would like to emphasize the social view examined in Schön's example, one that states that ordinary people working together can come up with answers to these questions. To give another example, in May 1968 the U.S. submarine *Scorpion* disappeared in the North Atlantic. The Navy knew its last reported location, but no more than this. As a result, the area in which it might have been found was 20 miles wide and, of course, many thousands of feet deep. Normally, of course, the solution to this problem would be to get a small number of experts who would give their expert judgment on where the submarine was.

However, a naval officer named John Craven had a different idea [see 22]. Instead, he constructed a series of scenarios about what might have happened to the *Scorpion* and gave these to a wide range of people with various forms of expertise: mathematicians, salvage operators, and so on. All these people were then asked to guess what happened to the *Scorpion*, and the Bayes formula (see Chapter 4) was then used to create an estimate of the location of the submarine. Eventually, the *Scorpion* was found 220 yards from where Craven's team thought it would be.

Now, it might seem like this is simply another kind of expert judgment except with more people, but the key point is that these were not experts in the TR sense. Instead, they were practical people with a wide range of expertise; in other words, they were *cognitively diverse* (to use the jargon). Another point is very important. Craven did *not* try to get the experts to agree with each other. This would, of course, merely produce a form of false consensus based on authority. Instead, he took the average of all the different guesses [22].

Now, this fits with a growing body of work that shows that cognitively diverse, nonhierarchical groups can produce better guesses than traditional (lone) expert judgment. For example, the sociologist Hazel Knight carried out a series of experiments in which she asked the students in her class to estimate the temperature of their room. She then took the average of the guesses. The average was 72.4°F; the room temperature was actually 72°F [23]. Other experiments produced similar results; the most famous is the beans-in-the-jar experiment, the key point of which is this: when a group is asked to estimate how many beans are in a jar, they invariably guess better (i.e., their average guess is better) than the vast majority of the guesses of *any of the individuals in the group* [24].

Note that this is radically different in terms of a conception of expertise than the traditional one. The traditional expert is a loner: he (and it is usually a he) uses rationality to work out the one true answer, as we have seen. But, in the *Scorpion* example and the beans example, it was *ordinary people* who did this. Moreover, there was discussion, sure, but no hierarchy; one person did not "call the shots." Instead, by taking the average, this was avoided (cf. the parallels with reliability trials in Chapter 3). The group was the unit of analysis, not the individual, but the group had to be *cognitively diverse*.

So, to apply this to what we were discussing, the Cali hospital case was an example of just this: taking the average (or median) of what everyone thought at the hospital. This involved talking to the real experts: the people who actually did the work. By reflecting on their practice, they then managed to self-regulate the system, and the accident rate decreased to its "natural" rate. In other words, the system *self-organized from the bottom up*. There was no leader or overall authoritarian manager at the top. Instead, the reflective practitioner enabled the organization to self-organize until it found its natural level of accidents. After this, the purpose of the reflective safety practitioner would be to continue to keep these processes and procedures in operation so that the organization could continue to self-organize, continue to be aware of its capabilities, via ensuring that information flowed from top to bottom and then from bottom to top. Information flow is the key.

8.8   WHY HAS THE ACCIDENT RATE GONE DOWN?

Now, all this discussion, in fact the whole book, leads us to an important question. We are arguing that something has gone wrong with safety science, specifically since the late 1960s (with the introduction of cognitivism) but more generally since the development of the science in the work of W. H. Heinrich. We are constantly told that the accident rate, for example in motorized transport, has gone *down* over this period. How do we explain this? Surely, this shows that our theories are wrong. The medicine appears to be *working*.

There is one answer to this question that is associated most with John Adams in his book, *Risk*. On page 60, he provided a series of figures showing mortality ratios for accidental and violent deaths for some 31 countries worldwide for the first 75 years of the 20th century. Adams argued that, apart from obvious blips (the graph zooms up in World Wars I and II), the rate stays fairly steady during the entire century. Adams explained this by his theory of risk compensation. Adams argued that we all have an inbuilt desire for risk, and that if safety measures become too stringent, then we will simply find a way to circumvent them and increase the risk we experience. Needless to say, this has become a controversial hypothesis, and we have no intention of joining in this heated debate. Adams concluded that there is "little to show" in the statistics for "all the labours of the risk reducers — the regulators ... the safety engineers and all the others involved in the safety industry over many decades" [25].

This is an extremely controversial viewpoint, and we merely note it and continue for the meantime. However, another aspect of his theory almost certainly has an element of truth in it. This is the idea that safety costs money, and that the more safety improvements that are introduced into any specific area (e.g., a mode of transport), the more expensive it becomes to use. People therefore choose to use cheaper transport, which means higher risks. So, benefits in safety are balanced, and accident rates are not directly linked to safety directives or initiatives as one might imagine.

8.9   INTERPRETING ACCIDENT STATISTICS

This still avoids the basic question, why do accident rates decrease even in specific industries or areas if the general approach to safety management is flawed (as we believe)? We have two answers to this question that tie in with the basic themes of this book The latter actually answers the question with which we began: why taxonomies and safety recording systems?

Our first answer is simple. To take an example from motoring, in 1949 R. J. Smeed plotted the relationship between fatalities per vehicle and vehicles per capita [27]. He discovered a law (which, importantly, has been supported by many other studies in other contexts). What Smeed found was that the more driving there is, the fewer accidents there are (per mile driven). In other words, the more cars on the road, the safer you are. This holds true today. In developing countries (in which, statistically speaking, it is highly risky to drive), there are far fewer cars on the road than there are in the United States or in Britain. This might seem to be counterintuitive, but that is because we are used to thinking of road accidents as purely caused *by motor vehicles*. (In fact, there is evidence that accident rates were, in some places, far higher when most transportation was with horse and cart [26]. This became even worse in the early years of the century when road transportation was a mix of horse-drawn carts and motor vehicles.) Moreover, in the developing world there is less industrialization, fewer cities, and fewer roads. This might be explained in cultural terms were it not that in Britain and the United States, in the early days of motoring, accident rates were similarly horrendous [27].

What seems to happen is that the more vehicles on the road, the more drivers are *forced* to drive safely. Not only can they not build up the requisite amount of speed (e.g., in a traffic jam) to have a serious accident, but also the sheer presence of large numbers of motor vehicles around you (who *regulate* your driving, e.g., by shouting at you if you break basic rules of traffic) seems to produce a sort of enforced social consensus of safer driving.* Moreover, driving on empty roads has hidden risks that are not often thought about: highway hypnosis, the risk of animals and so forth jumping in front of the car, the risk of falling asleep, and so on. In the city, you are forced to pay attention for these things; on a deserted motorway in the middle of nowhere, you not only think you are safer, but also tend to be traveling at much faster speeds. If there is an accident, it is much more likely to lead to a fatality than in the center of a city at rush hour [28].

In other words, if we stop thinking about the psychology of individual drivers and start to look, as we ought to, at this as a systems issue, then we can see this is an example of the road-car-driver system self-regulating or *spontaneously generating order*. The infinitely large number of decisions of individual drivers have an emergent effect, which is that in the countries with the most cars, where there is the most driving, there are the fewest motoring fatalities (per mile traveled).

(Another reason for the decline in accidents might be that, in more rural areas, it is harder for emergency services to get to injured drivers or passengers quickly enough. It has also been suggested that better medical science and the increased speed with which emergency services can get to the victim might have helped to reduce *fatality* rates at least [i.e., injuries that would once have led to fatalities now merely become injuries] [29].)

Now, the key point about traffic is that there is no homunculus or central processing unit (CPU) for the traffic system, and no one has suggested that there should be one. Driving is an inherently decentralized activity because (noncommercial) drivers can drive where and how they want. The bureaucratic and organizational difficulties that stop safety improvements cannot arise in a system like the traffic system. The system self-organizes (over time) because it is able to do so.

However, in organizations the situation is very different. There is inherently a hierarchy and structure in an organization that there is not in (for example) the traffic system, and that is where the problems start. What we are arguing here is that

---

* As usual, we are talking statistically here, and we have to be very careful in terms of interpreting statistics. For example, whereas in cities there may well be fewer accidents, nevertheless there is a greater risk to pedestrians just because there are some. On motorways in the middle of nowhere, there are far fewer pedestrian casualties just because there are no pedestrians.

reporting systems and taxonomy theory can be a method for helping the system (in this case, the organization) self-organize.

It must be stressed that this is not just a pious hope. There is already evidence (from the chemical industry) that reporting systems similar to CIRAS and SECAS can actually cut accidents by as much as 75% (measured via lost-time injuries) over a 13-year time period [30].

8.9.1 Locus of Control, Learned Helplessness

However, there is another reason why accident rates (at least in specific circumstances) might have gone down. We have known for some time, for example, that one of the major predictors of propensity toward accidents is *social class* (although this is not something one tends to read about in the traditional literature) [31]. For example, child pedestrian accidents are five times higher in the lowest socioeconomic grouping than in the highest. Of course, this makes perfect sense: it is not, on the whole, managers and civil servants who have to face, for example, the industrial risks that one faces down in the pit or in a steelworks. Another major predictor of accidents is ethnicity (and gender), although one reads even less about that in esteemed journals [32, 33]. (Even if one did, it is not at all clear how these sociological data would relate to the primarily psychological approach preferred in contemporary safety science.)

We can go further than this by talking briefly about locus of control. This is yet another seeming diversion from our main theme, but it is the last one, and we hope it will tie in all the themes we have discussed.

8.9.1.1 Attribution Theory and Safety

Recent work has shown some evidence that people's propensity to risky behavior is related to the explanations they provide for accidents (we looked at attribution theory in Chapter 1). For example, in a study of South African taxi drivers, Peltzer and Renner found that participants who were more likely to produce superstitious causal explanations for accidents were more likely to state that they had an accident in the previous year than nonsuperstitious drivers [34]. These superstitious drivers were also likely to have a lower socioeconomic status than the nonsuperstitious drivers. In a similar study carried out in Nigeria, Dixey found that superstitious causal explanations for accidents were associated with fatalism: the belief that one's fate is not in one's own hands but is controlled by external forces (usually supernatural). She speculated that the prevalence of these beliefs may help explain Nigeria's (poor) health and safety record [35].

The key point is that these superstitious explanatory styles are associated with an external locus of control. In other words, the more external your locus of control is, the more likely you are to adopt an attitude of fatalism: what is the point of taking care of yourself, or trying to avoid accidents, fastening your seatbelt, and so on if it is all in the hands of fate or some similar external cause? In other words, it is not the superstitious discourse per se that is the problem, but the powerlessness that underlies it.

Interestingly, this brings us back to Leonard Sagan's point we discussed in Chapter 1. Sagan (in his book, *The Health of Nations*) argued that it was this attributional shift between an external and an internal locus of control that was the primary cause of the increase in longevity we have seen in the 20th century [36].* Moreover, his work has also been supported by Michael Marmot, who in his more recent book, *The Status Syndrome*, made fundamentally the same point [37]: it is one's locus of control (i.e., the more control one feels one has over one's life) that is the primary determinant of longevity, and this is strongly linked with social class and social status. In other words, to put it bluntly, if you feel yourself to be worthless, contemptible, or to have no control over your life, then this will be associated with a shorter lifespan. Sagan made the obvious link, that this means (as the Peltzer study indicated) that such attitudes will be associated with a greater propensity toward accidents as well. (This point becomes even stronger if one accepts the arguments of David Philips that many accidents are in fact suicides. The link among depression, suicide, and an external attributional bias are well known [38, 39].) This would help explain the differences in accident rates across social class that we have seen. Powerlessness and lack of control lead to an external attributional style correlated with propensity toward accidents.

Now, this is an interesting theory, but is it true? Apart from these two studies, we have two pieces of evidence that suggest it might be. There are a number of experiments and studies that indicated the importance of control (i.e., having an internal locus of control) in terms of job performance. Most important is the study by Edkins and Pollock, which indicated that job satisfaction (or lack of it) is correlated with passing red signals in train drivers. In other words, the happier you are in your job, the less likely you are to have a signal passed at danger [40].

We conducted an interesting study in this area ourselves using the CIRAS database. We attempted to discover the mental health of the organization in terms of its attributional biases. We discovered that the classic depressive attributional style was associated with fatigue in the same database. Of course, fatigue was also associated with other safety conditions. So, it seems that our theory is correct. Therefore, one of the major aims of any safety manager should be, via reporting systems, to empower the workforce and attempt to restore their locus of control inward, to make them enjoy their jobs and feel valued. This, we would argue, would have as much, or more, effect on accident rates than more conventional approaches [41].

In response to the question, why have accident rates gone down? we think there are three answers, one trivial and two profound. The trivial answer is simply that medical services, emergency room technology, and the increased ability of emergency crews actually to get to the scene of the accident have reduced accident rates. Second, given that society, like all complex biological systems, has a propensity to self-organize and achieve the shifting homeostasis, our own society has spontaneously moved toward reducing its own accident rates. Smeed's law is a classic example of this. After all the safety improvements that have been made (e.g., seatbelts, air bags, etc.), it is at least arguable that it is the raw fact of the *number of cars on the road* that is the best predictor of accident rates. The greater the number of cars, the greater this impacts (no pun intended) on other drivers' driving styles,

---

* Sagan was discussed in Lowentin's explanations of the drop in the tuberculosis death rate.

slowing them down, making them more cautious, and lowering the accident rate. In other words, the transport system has, to a great extent, self-organized.

The final answer is this: There may well be group psychological reasons for accidents, unnoticeable (or barely noticeable) in the individual, but that become progressively more obvious as we look at more people. These are related to power, autonomy, and control and manifest themselves through (among other things) attributional styles. The richer someone is, the more control the person has over his or her life, the more autonomy, the more power. His or her attributional style tends to be more internal. As the countries in the West have become richer, the attributional style has tended to move inward [42]. We have given some evidence (not conclusive, but intriguing) that this may well be related to propensity to have accidents, again not noticeable at the level of the individual but discernable when one looks at statistical averages. It helps to explain why, even in our own wealthy societies, social class is still a good predictor of accident rates: the poorer you are, the more powerless you are, the less autonomy you have, the more external your attributional style is, and so on.

8.10 EMPOWERMENT

This chapter has covered a lot of ground very quickly, but perhaps it would make things clearer if we brought back a concept from psychology that we discussed briefly in Chapter 5: connectionism. Connectionism is, as we have stated, an information-processing theory, but it is still useful to discuss it here as a metaphorical paradigm because it is a very different information-processing theory from cognitivism. Connectionism posits cognition as produced by individual neurons in the brain and their relationship to one another, so computer models of this process use nodes (computer models of neurons) and the extent to which these interrelate. In connectionist models, cognition arises from a bottom-up, emergent, and distributed process that is highly contingent on *experience*. As the brain meets more and more different experiences, more and more links are made between the neurons to build up a web of experiences; crucially, no rules are posited as existing in the brain before this process begins.

This is radically different from the cognitivist approach, which, you will recall, is instead a top-down approach. In cognitivism, there is always a leader or controller at the top of the system, which might be termed a homunculus or a CPU or something of that sort. This controller creates, passes down, and enforces these rules. In most cognitivist models, these rules are genetically preprogrammed in the brain. This model is hierarchical and posits cognition in the shape of a pyramid, with the CPU (i.e., a computerized homunculus) at the top and more and more subsystems as one goes down, ending with individual neurons flicking on and off.

We hope we will not be seen as guilty of overstretching this metaphorical dichotomy if we view these two views as a symbolic link between our views of science, psychology, and the organization. In all three cases, there are two possible views. The first general view goes like this: in science, it is the realist/Platonic position with theoretical physics at the top; in psychology, it is cognitivism; and in the organization, it is the standard rule book managerial approach. In all three situations, the basic model is that of a pyramid, with physics, the CPU, or management at the top, creating rules or laws that are then passed down by (implicitly) dumber and dumber subgroups!

Instead of this view, in all three situations we are stressing the importance of decentralization: in science, emergence, distributed systems, and bottom-up processes; in psychology, a connectionist approach as opposed to a cognitivist approach; in organizations, the patchwork quilt, not the pyramid. We hope this metaphorical link makes it clear what type of principles and practices we are suggesting for safety management practice.

Finally, therefore, this brings us to empowerment in organization practice. This is a word that has been given a bad name in recent years because some companies have used it as a smokescreen to hide behind instead of giving *genuine* empowerment strategies [43]. We do not hold out much hope for a simple use of consultancy buzz words like *stakeholder* and *ownership*. However, what we are arguing here is that *genuine* empowerment strategies might be a way forward for improving safety.

We would argue that there is evidence available that empowerment/decentralization helps safety. For example, studies of the mining industry have demonstrated that there is a strong correlation between size of coal mine and accident rate: the accident rate is nearly twice as high in large mines as opposed to smaller mines. There are numerous other studies that found similar results. The best way of keeping the size of the working unit down is via *decentralization*. There is something of the virtuous circle going on here because decentralization also "reduces communication problems, makes participation in decisions easier and increases job satisfaction" [44], all of which tend to improve safety.

There are other studies that back up the importance of empowerment, decentralization, and participation by staff in decision making. For example, work has been done in China to create empowered workers via worker participation in health and safety committees. This has led to the identification of many safety issues of which management was previously unaware [45], a classic example of the people at the front end knowing more about the true safety situation than those further up the management structure. (See also [46] for an empowerment project that reduced accident-related absence by 74%.) These studies took place in industrial settings, but examples in empowerment zones in the United States implied that similar improvements in public health can occur in a civic setting [47]. The key point would be to prioritize bottom-up initiatives that emphasize participants taking control of their own lives rather than top-down, rule-based initiatives.

The key point about empowerment is that it moves the attributional locus of control from the external to the internal. You, quite literally, take control over your own life, your own work practice. As you gain autonomy (control), this constitutes yet another factor to create the virtuous circle of helping the man-machine circle of the organization to self-organize.

8.11 CONCLUSION

Finally, we would argue that we should replace the pyramid model of both science and organizations, with its emphasis on rules, with a patchwork quilt model based on cases. Instead of being top down, we would see the organization as bottom up. The workforce is the expert. We, as safety managers, are facilitators helping them reflect on their practice. In other words, we can now answer the questions (asked in Chapter 1) of whither safety management and safety science. We would answer thus: the purpose of safety science is to help the organization self-organize to reach whatever level of safety is the right one for that particular organization. Safety science is the development of techniques to facilitate this.

We view taxonomy theory as a means to this end. On the macrolevel, it is a methodology that enables meaningful reliable data to be passed upward to managers so this information might be acted on, and these actions passed down, with the staff's reactions to these actions then passed back up again, and so on. It is hoped, therefore, that a virtuous circle can be created in which both staff and managers benefit. As the staff are empowered, then they take responsibility for their *own* safety practice. Then, it is hoped that their locus of control (which can be analyzed via analysis of the discourse used in the actual reporting system) demonstrates a move toward the internal. This again will lead to a reduction in the accident rate, until this reaches its natural rate (which will, presumably, be unique to each organization).*

So, taxonomy theory works by helping systems self-organize and helping to empower staff create their own solutions to safety issues. Even at the microlevel (i.e., the way a reliability trial is carried out), we view it as depending on the *social consensus on events*, but not in a top-down hierarchical way. Instead, it is a way of getting people to cooperate and creating real consensus, *not* a sort of Orwellian consensus in which (however covertly) the boss tells people what to think, and then they agree with him. Instead, by the mechanics of the reliability trial (in which no one view is privileged but instead a consensus is allowed to emerge from the group), shared social meanings and interpretations are built up, which then act as the basis for later interpretations of later reports or cases. So, on both the macro- and the microlevels, groups (made up of individuals with viewpoints and presuppositions) self-organize and create their own meanings and their own interpretations and thereby *empower themselves*. They thereby make their own practice, and that of the organization in which they work, safer.

This is where we came in.

REFERENCES

1. For an example of instrumentalism, see Quine, W., Two dogmas of empiricism, *Philosophical Review,* 60(1), 20, 1951.
2. Harris, R., *The Linguistics Wars,* Oxford University Press, Oxford, U.K., 1995.
3. Wittgenstein, L., *Philosophical Investigations,* Blackwell, London, 1999, p. 81, section 201.
4. Schank, R. and Abelson, R., *Scripts, Plans Goals and Understanding,* Erlbaum, Hillsdale, NJ, 1977, p. 67.
5. Klein, G. and Calderwood, R., How do people use analogies? in Kolodner, J. (Ed.), *Case-Based Reasoning: Proceedings of a Workshop on Case-Based Reasoning,* May 10–13, Clearwater Beach, Florida; Morgan Kaufman, San Mateo, CA, 1988, p. 210.

---

* To repeat a point, there is empirical evidence that this is in fact what happens in practice.

6. McKinney, E. and Davis, K., The effect of deliberate practice on crisis decision performance, *Human Factors,* 45(3), 436, 2003.

7. Woolfson, C., Foster, J., and Beck, M., *Paying for the Piper,* Mansell, London, 1997, pp. 334–336.

8. Semler, R., *Maverick,* Arrow, London, 1994, pp. 92–96.

9. Strauss, A., Schatzman, L., Ehrlich, D., Bucher, R., and Sabshin, M., *The Hospital and Its Negotiated Order,* Macmillan, London, 1963, p. 147.

10. Polanyi, M., *Personal Knowledge: Toward a Post-Critical Philosophy,* University of Chicago Press, Chicago, p. 1958.

11. Waldrop, M., *Complexity,* Simon & Schuster, New York, 1992.

12. Cartwright, N., *The Dappled World,* Cambridge University Press, Cambridge, U.K., 1999.

13. Popper, K., *The Poverty of Historicism,* Routledge, London, 1961.

14. Bandura, A., *Social Learning Theory,* General Learning Press, New York, 1977.

15. Schön, D.A., *The Reflective Practitioner,* Basic Books, London, 1991.

16. Heinrich, H.W., *Industrial Accident Prevention,* 4th ed., McGraw-Hill, New York, 1959.

17. Reason, J., *Managing the Rights of Organizational Accidents,* Ashgate, London, 1997.

18. Rasmussen, J., *Information Processing and Human/Machine Interaction,* Elsevier, Amsterdam, 1986.

19. Schön, D.A., *The Reflective Practitioner,* Basic Books, London, 1991, p. 191.

20. Miller, C., The presumptions of expertise: the role of ethos in risk analysis, *Configurations,* 11, 169, 2003.

21. Miller, C., The presumptions of expertise: the role of ethos in risk analysis, *Configurations,* 11, 169, 2003.

22. Sontag, S. and Drew, C., *Blind Man's Bluff,* Public Affairs, New York, 1998, p. 146.

23. Knight, H., A Comparison of the Reliability of Group and Individual Judgments, master's thesis, Columbia University, New York, 1921.

24. Traynor, J., Market efficiency and the bean jar experiment, *Financial Analysts Journal,* 47, 43, 1987.

25. Adams, J., *Risk,* UCL Press, London, 1995, pp. 60–61.

26. Leeming, J., *Road Accidents,* Cassell, London, 1969.

27. Adams, J., Smeed's law: Some further thoughts, *Traffic Engineering and Control,* 28(2), 70, 1987.

28. Tenner, E., *Why Things Bite Back,* Fourth Estate, London, 1996, p. 267.

29. Reinhardt-Rutland, A., Has safety engineering on the roads worked? Comparing mortality on road and rail, in McCabe, P. (Ed.), *Contemporary Ergonomics 2003,* Taylor & Francis, London, 2003, p. 341.

30. Jones, S., Kirchsteiger, C., and Bjerke, W., The importance of near miss reporting to further improve safety performance, *Journal of Loss Prevention in the Process Industries,* 12, 59, 1999.

31. Roberts, I. and Power, C., Does the decline in child injury vary by social class? *British Medical Journal,* 313, 784, 1996. See also Roberts, I., Cause specific social class mortality differentials for child injury and poisoning in England and Wales, *Journal of Epidemiological Community Health,* 51, 1997, 334.

32. Weddle, M.G., Bissell, R.A., and Shesser, R., Working women at risk. Results from a survey of Hispanic injury patients, *Journal of Occupational Medicine,* 35(7), 712, 1993.

33. McCracken, S., Feyer, A.M., Langley, J., Broughton, J., and Sporle, A., Maori work-related fatal injury 1985–1994, *New Zealand Medical Journal,* 114(1139), 395, 2001.

34. Peltzer, K. and Renner, W., Superstition, risk-taking and risk perception of accidents among South African taxi drivers, *Accident Analysis and Prevention,* 35, 619, 2003.

35. Dixey, R.A., "Fatalism," accident causation and prevention: Issues for health promotion from an exploratory study in a Yoruba town, Nigeria, *Health Education Research, Theory and Practice,* 14(2), 197, 1999.

36. Sagan, L., *The Health of Nations,* Basic Books, New York, 1987.

37. Marmot, M., *The Status Syndrome,* Bloomsbury, London, 2004.

38. Philips, D., Motor vehicle fatalities increase just after a publicised suicide story, *Science,* 196, 1464, 1977.

39. Cialdini, R.B., *Influence,* Allyn & Bacon, Boston, 2001.

40. Edkins, G. and Pollock, C., The influence of sustained attention on railway accidents, *Accident Analysis and Prevention,* 29, 533, 1997.

41. Davies, J., Ross, A., Wallace, B., and Wright, L., *Safety Management: A Qualitative Systems Approach,* Taylor & Francis, London, 2003.

42. Lerner, M., *The Belief in a Just World: A Fundamental Delusion,* Plenum, New York, 1980.

43. Argyris, C., Empowerment: The emperor's new clothes, *Harvard Business Review,* 76(3), 98, 1998.

44. Argyle, M., *The Social Psychology of Work,* Penguin, London, 1981, p. 199.

45. China Capacity Building Project–Project Co-ordinating Committee, *Final Report of the Project Co-ordinating Committee,* http://ist socrates.berkeley.edu/~lohp/graphics/pdf/CHINARPT.pdf, May 29, 2002.

46. International Survey Research, *Case Study: Using Cooperation and Empowerment to Improve Safety Records* (http://www.isrsurveys.com/pdf/insight/CaseStudy101.PDF), 2005.

47. Stoker, P., Who is empowered? Innovative governance in Baltimore's empowerment zone, in Harris, M. and Kinney R. (Eds.), *Innovation and Entrepreneurship in State and Local Government,* Lexington, Lanham, MD, 2003, pp. 99–119.

9 Conclusion

It should be clear by now that, as we said at the beginning of the first chapter, although we are concerned with the same issues as most safety experts (reducing accidents, promoting safety), our approach is different (some would say radically different) from the standard view in the field. In this concluding chapter, we summarize what we have been saying and look at some of the broader implications of our approach.

If there is one theme that runs through this book, it is our belief in bottom-up concepts of self-organization in which people and organizations are allowed to solve their own problems, as opposed to top-down solutions in which an elite *imposes* a solution on the problem, usually in the form of rules and regulations. And, taxonomy theory, despite that we consider it innovative in its own right, should always be seen as merely a means to a self-regulatory system. Although much of our work remains commercial-in-confidence, we have tried to be specific when appropriate, describing ways of facilitating information flow.

Information must flow in any organization, and that means information must flow not only between groups and from management to staff, but also (this is often forgotten) from staff to management. It is not enough for this information simply to stay static, locked away in filing cabinets or on computer hard disks. The information must be dynamic: it must flow constantly, in all directions throughout the organization. Even more, this information must be acted on.

The impression we gained from the Confidential Incident Reporting and Analysis System (CIRAS) was that there was an inverse relationship between the amount of money spent on safety and the subjective impression the staff tended to have about how seriously that company took safety. To spell it out even more plainly, the more the company spent, the less the staff liked it.

This strikes most people as so wildly anti-intuitive that they are inclined to doubt it until they hear the reasons given for this, at which point it seems to make more sense. The fact is that most money spent on behavioral safety is spent on training courses, information campaigns (mainly of the remember-to-be-safe variety), and introducing new procedures (rules). In other words, most of the money is spent on information flow from management *down* to staff. Advertising campaigns and training seminars are unidirectional. There are no training seminars *for* senior management held *by* staff, in which the managers are lectured (in a vaguely patronizing fashion) about how seriously they ought to take safety and about the need to avoid human error. One might almost imagine managers not to be human. Despite the rhetoric of safety first, the rigid structure of the information flow emphasizes that the workers' role in these things is to remain passive while managers (specifically, in this instance, safety managers) *tell them* what to do, how to behave, and what to say.

There is little evidence that this approach makes things better. Actually, it might make them worse by further alienating the workforce and reducing job satisfaction. We have noted that decreased job satisfaction has been directly linked to certain accidents, for example, signals passed at danger (SPADs) in the railways [1].

The alternative to this approach is a strategy of empowerment. Now, we are aware the word has something of a bad reputation in current management studies, and there is a reason for this. Far too frequently, managers have offered the illusion or appearance of empowerment without the reality, and the crux of the issue is trust [2]. Trust is now seen as one of the key features of an effective and meaningful managerial strategy. However, trust cuts both ways. It is certainly true that staff should trust their managers and that managers should keep their promises, be seen as fair, and so on. But, the corollary of this is that managers should also trust their staff and should act on the assumption that the staff (who, remember, have been hired in the first place by management) are *not* going to commit rule violations (without a good reason) and are *not* going to be incompetent, stupid, or worthless unless there is a good reason to think otherwise. In other words, staff should be allowed to get on with doing their jobs, a view summarized in the wise maxim that "the manager who manages best manages least." The manager should have faith in the staff's ability to organize because, as we have pointed out, an organization is a system, and systems do spontaneously self-organize unless top-down initiatives strangle that tendency and put a stop to the natural creativity that staff generally manifest when left on their own.

Now, we are obviously biased (but then, so is everyone). However, we do genuinely believe that reporting systems and databases of the kind discussed in this book can be a key factor in allowing information to flow in the ways we have described. In theory, almost any reporting system that gathers information from staff is better than none. But, we would argue that to create a genuinely effective system some form of hierarchical taxonomy is necessary so that data can be analyzed quickly and easily. We have also argued that a reliability or consensus trial is the only way of ensuring that such systems are meaningful.

However, in the jargon, reliability does not ensure validity. Given the epistemology we have come to adopt, the use of these words is problematic. However, in the broadest sense of the word, we are arguing here that validity simply means whether managers actually act on the information they are given. Nothing, in our experience, will undermine a reporting system more quickly than the view that the information gathered is merely gathered for form's sake, and that no concrete actions are actually taking place as a result. Lack of action can sometimes result from managers becoming defensive on discovering (and this is especially true with a confidential system) that not all the staff have such a high opinion of the manager's abilities as they do. To discuss how to deal with this issue is outside the scope of this book because it deals with the broader issue of how to get beyond an adversarial culture and develop one of genuine cooperation. Nevertheless, we would say that, in our experience, once real action is taken through information gathered, then a virtuous circle can be created in which staff respect managers for acting on their safety concerns, and as a result of this, morale and motivation increase, which in itself can lead to a decrease in accidents. We are not discounting the difficulties that stand in the way of such a goal, but it is surely a worthy aim.

9.1 SCIENCE, ETC.

However, the implications of this book clearly go beyond safety management. Although we began this work with pure safety management in mind, we must emphasize that we now have a broader outlook and see these issues as having a far wider application throughout the safety and risk sciences (and beyond). For example, one of the authors (B. W.) has recently begun working with a children's charity in the United Kingdom on issues relating to self-harm and suicide. The issues in this context, although they might seem a million miles away from nuclear/train safety, are essentially the same. This children's charity (like the nuclear industry) has a statutory requirement to list all accidents and incidents that take place among the children in their residential care homes. These incidents and accidents are logged in textual (qualitative) form and then coded and put in a centralized database. Moreover, for children who are perceived to be particularly at risk, a risk assessment is carried out on their behavior.

It is fairly obvious that the basic procedures (logging unwanted events) and aims (reducing incidents and accidents) are almost exactly the same in this children's charity as they are in many high-consequence industries. We thus feel they are best addressed within the framework of the taxonomic work we have presented in this book. Specific issues that might ring a bell given what we have discussed in this book include different definitions of self-harm across residential homes (i.e., a lack of reliability); a different set of procedures regarding what to do in the event of self-harm; differing interpretations of risk; and so forth.

Another recent example we have encountered is a lack of consensus in the use of certain classifications in police databases (e.g., a lack of a clear definition for the category persistent offender). All of these issues, we would argue, can only be addressed via openness in classification, followed by reliability testing of the sort that we describe in Chapter 3 and the appendix.

But, the implications of addressing taxonomic issues go even further. They cut to the core of what we mean by the words *science* and *expert*. More specifically, they cast doubt on the difference between the hard sciences and the soft (or human) sciences, especially the way this difference represents itself in psychology/sociology: as the difference between quantitative and qualitative data. We have argued (in Chapter 2) that the ontological distinction between words and numbers is hard to sustain. What matters is simply that classification systems (numbers are a good example) are reliable. Moreover, these classifications are neither programmed into the human brain nor found floating in the metaphysical ether. Instead, they are the result, primarily, of social negotiation in a pragmatic context. Human beings are active. They carry out tasks. As a result, they frequently have to categorize things such that other human beings can understand and use the classifications. If we could not do this, social life would break down. Numbers are just another taxonomic tool, as is the taxonomy: quantitative/qualitative.

Once this is understood, then of course we realize that all the categorizations that we use have lost their metaphysical nature. We categorize the world in a certain way, but this is not necessary. We can *choose* how to categorize things. This might give some people a sense of metaphysical vertigo, but it clearly also opens up new horizons and gives us a sense of freedom to solve problems that otherwise might have seemed intractable.

This brings us back to the qualitative/quantitative divide. We have chosen to talk only about safety-related texts in Chapters 2–4, but we could equally be classifying any kind of text (or any kind of object, for that matter). Data produced from this process, in the form of coding frequencies, can be treated as quantitative in the strictest sense of that word. Quantitative data are *always* generated from socially agreed taxonomic decisions, and given that this is the case, there is no ontological reason to treat coding frequencies from textual observation any differently from height frequencies from human being observation, or accident frequencies from train observation, or any other form of quantitative data.

This offers the possibility of true integration of data sources in psychology and sociology as well as in safety science. This in turn creates the possibility of a whole new world of research, one that we ourselves have hardly even begun to explore. Among other things, viewing all distinctions as taxonomic moves us beyond the cognitivist dualisms of individual/environment and language/behavior to a world where patterns of discourse and other behaviors (in staff or other social groups) can be studied (modeled) in their own terms to see what emergent properties they may display. The CIRAS database was, among other things, a dynamic database of the discourse of the British rail industry that could be tracked for years and integrated with external phenomena (prominent rail crashes, changes in the socioeconomic setup of the industry) to model the entire socio-technical-political system. This, we would argue, is a fundamentally more reliable and valid approach than the standard cause-effect approaches like conducting a root cause analysis or handing out some questionnaires and then using frequentist statistical techniques (factor analysis, regression, etc.) to test hypotheses.

In line with our philosophy of emergence and self-organization, what we are particularly interested in is themes that emerge from the data, not themes we impose on them. In other words, as we stated, we look *at* the data and do not attempt to look through it. Moreover, the temptations of pseudo-objectivity should be resisted. We are specific human beings looking at data in a particular context, with particular biases and assumptions. It is simply not scientific to pretend any different. And, this means to adopt (in the broadest sense of the word) a Bayesian viewpoint that acknowledges these facts and does not attempt to brush them under the carpet because this ought not be the case.

This is an exciting time to be a social scientist. In psychology, the information-processing paradigm has crumbled and is now replaced by the new psychologies, which we discussed in Chapter 5, a view that promises fundamentally new and profound insights into what it means to be human. Moreover, in statistics, Bayesian approaches are becoming more and more popular as the assumptions of null hypothesis testing come under increasing attack. However, what we have tried to show in this book is that these phenomena are really two sides of the same coin: what they indicate is the coming collapse of the technically rational view of science (at least in the social sciences), which was useful once but is now actively holding up progress. We hope that taxonomic theory and thinking will prove to be an important link in the chain of the new methodologies and techniques that will have to be developed to create new social sciences for the 21st century.

REFERENCES

1. Edkins, G. and Pollack, C., The influence of sustained attention on railway accidents, *Accident Analysis and Prevention,* 29(4), 533, 1997.
2. Kramer, R.M. and Cook, K., *Trust and Distrust in Organizations,* Russell Sage, New York, 2004.

Appendix: Carrying Out a Reliability Trial

This appendix is intended to help those considering carrying out an interrater reliability trial.* There are some important things to keep in mind from the outset.

First, it should be made explicit at all times that *the thing tested is the taxonomy not the coders*. Not only is this important philosophically (because otherwise we get mired in misconceptions about "right" and "wrong" codes; see Chapter 2 on the privileged classifier), but also it helps to get coders to engage with coding trials and give their own honest interpretations. This is the only way to find out how a taxonomy will work in practice.

Second, at the point of coding, coders should be physically separated to give a "fair trial." Otherwise, well-intentioned discussions tend to bias coding in particular directions. Granted, this social coding will sometimes be more like the way codes will be assigned in real life (when people are bound to discuss events), but it cannot tell us whether information derived is useful because this can only be the case if coders agree on coding events when they do this blind to other coders' decisions. Whether codes assigned to our databases are independent of individual coder bias is, after all, what we are trying to test.

In rough chronological order, the process for a taxonomy trial for interrater reliability is as follows:

- We need to decide who should be involved in a trial. Pragmatically, it often makes the most sense to involve people who will be using the system in the "real world" in trials. If two or three people are likely to be involved in using the taxonomy, then all should be asked to code events for the trial. If there are many more potential (or actual) coders, some "sampling" will be required. Issues to consider may be coders with different degrees of training or experience or different backgrounds or expertise. We would recommend six to eight coders as a good number for a trial. Greater numbers involve calculations that can be prohibitively time consuming. If eight coders code,

---

* Intrarater trials involve most of the same principles that are outlined here (coders blind to comparison codes; independent marking; discussion and feedback), but a separate concern with an intrarater trial is how long the time lag between tests should be. This will usually depend on how many events coders have processed in the interval. If the time lag is too short (i.e., not enough events coded in between), coders will often not reinterpret events but will simply recognize them and remember their initial codes. We do not really want to test recognition; rather, we want to know how an event coded one way in the past would *now* be processed. So, we need to wait long enough for events to "appear as new."

say, 20 events and are allowed 3 codes for each, then there are 480 individual codes for analysis; 560 paired codes for comparison (giving 28 average 'raw' agreement scores); 112 different sensitivities and specificities; and so on, which should give enough information on the system.

- It is fine to fully inform coders in advance about the purposes of the trial. This can be done in a group with everyone present. Tell them they are to try to code events into the taxonomy. Ask them about their understanding of the categories. Let them discuss any different interpretations that may emerge so they can have a fair attempt at using the codes.

- Next, the cases or events to be processed have to be chosen. These can be a broad selection that you intend to cover different codes, can be picked at random, can be picked to illuminate specific coding decisions, or simply can be taken as a block from a given time period. Selection criteria should be made explicit when reporting. How many events are used is usually decided on pragmatic grounds, depending on how long and complex the texts are, but obviously, the more codes that are assigned during the trial, the more information there will be available afterward. Single "day trials" with simultaneous coding are obviously limited by the number of events that can be processed during the time available. If events are frequent and the taxonomy is up and working, getting coders to "double code" a small percentage of events (e.g., to code 10% of events another coder has coded) on an ongoing basis for a short time can yield good data. However, in these less-formal trials, the strict condition of no consultation during the trial (see below) can be harder to manage.

- Wallace et al. [1] noted how reliability tends to be more than 10% better when specific items to be coded are presented rather than presenting more general texts from which they have to decipher what to code. For example, reliability of a human factors taxonomy can be increased if the person holding the trial first decides what the events to be coded are and limits the trial to classification of these specific issues. If, however, coders have to read a full event investigation or safety report (or observe or investigate an event themselves), then they have two tasks: first to pull out the human factors, then to classify them. *Reliability will be a function of both of these tasks.* Trying out both situations is probably best. Then, you can work out the relative levels of problems of recognizing what to code and issues with actual processing of events to be coded through the taxonomy. If you preselect "chunks" from investigations and, say, code this event as a type of error, then you should be aiming for a reliability of 80% or above, so that there is room for some uncertainty regarding whether different people will choose the same features to code in practice.

- Whichever method is chosen, trials should be carried out under strict conditions. There should always be a trial supervisor or organizer: separate coders should process the same set of examples in separate rooms; no consultation between coders should be allowed during the trial itself, although the organizer may be on hand to answer specific queries. If the trial is an ongoing one (see above), supervision by an independent person is especially important.

- Some systems will be electronic (computerized), and others will involve "tick box" coding sheets. As we are interested in the taxonomic health of the database, it may often be useful simply to present texts and codes on paper. This way, misalignment of coding because of different degrees of "mouse dexterity" or familiarity with electronic systems is controlled. We are interested in interpretations more than practical skills at this point. Granted, some electronic systems may *improve* reliability (e.g., by allowing for the presentation of codes in a hierarchical fashion, one logic gate at a time), but if the system is robust, it should stand the pen-and-paper test. Pen-and-paper tests may also be the most practical if we have multiple coders and want them to code simultaneously in different rooms.

- For the purposes of analysis, classifications should be collected, and "marking" or calculation (see Chapter 3) should be done by the supervisor or another person not involved in the trial.

- Finally, feedback should be given to and obtained from all coders. Biases and disagreements should be pointed out and discussion encouraged regarding how they might have occurred. It is vital to obtain responses from coders concerning their thought processes when interpreting contentious cases. This allows for taxonomic distinctions to be tightened or redesigned because conceptual overlap in the coding choices is eliminated (see Chapter 3). Obviously, if there is high reliability between a group of coders and another coder is clearly out of line, then this information can also be useful, and the approach here would involve managing the person rather than the taxonomy. Most often, in our experience, high reliability is achieved through a combination of logic work (with the taxonomy) and social work (with the coders).

- There is no strict guide regarding how often trials should be conducted, but ongoing trials are useful if resources allow, and trials are necessary when new systems are introduced (or systems have been altered) and when new coders begin to use a system. They may also be prompted by concerns about reliability. For example, when the frequency of a certain type of event increases markedly from one time period to the next, we want to know whether this is evidence for a shift in terms of our organizational system (i.e., is reliable evidence) or merely shows a realignment of our taxonomic decision making.

REFERENCE

1. Wallace, B., Ross, A.J., Davies, J.B., Wright, L., and White, M., The creation of a new Minor Event Coding Scheme, *Cognition, Technology, and Work,* 4(1), 6, 2002.

Related Titles

Accident/Incident Prevention Techniques
Charles D. Reese, University of Connecticut, Storrs, Connecticut
ISBN: 0415250196

Human Error: Causes and Control
George A. Peters, Peters & Peters, Santa Monica, California
Barbara J. Peters, Peters & Peters, Santa Monica, California
ISBN: 0849382130

Human Safety and Risk Management, Second Edition
Ian Glendon, Griffith University, Queensland, Australia
Eugene McKenna, University of East London, London, United Kingdom
Sharon Clarke, University of Manchester, Manchester, United Kingdom
ISBN: 0849330904

Increasing Productivity and Profit through Health and Safety: The Financial Returns from a Safe Working Environment
Maurice S. Oxenburgh
Penelope S.P. Marlow, Pepe Marlow & Associates, Concord NSW, Australia
Andrew Oxenburgh, Consultant, Clapham, United Kingdom
ISBN: 0415243319

Prevention of Accidents through Experience Feedback
Urban Kjellén, Norwegian University of Science and Technology, Sentralbygg, Norway
ISBN: 0748409254

Slip and Fall Prevention: A Practical Handbook
Steven Di Pilla, ESIS, Inc., Haddon Heights, New Jersey
ISBN: 1566706599

Index

Printed and bound by CPI Group (UK) Ltd, Croydon, CR0 4YY

23/10/2024

01778242-0012